Cosmic Memory

PREHISTORY OF EARTH AND MAN

By

Rudolf Steiner

Translated from the German by Karl E. Zimmer

1817

HARPER & ROW, PUBLISHERS, San Francisco
Cambridge, Hagerstown, Philadelphia, New York
London, Mexico City, São Paulo, Sydney

The authorized translation for the Western Hemisphere by agreement with the Rudolf Steiner-Nachlassverwaltung, Dornach, Switzerland.

This book is published in Switzerland under the title *Aus der Akasha-Chronik.*

FIRST HARPER & ROW PAPERBACK EDITION PUBLISHED IN 1981

Library of Congress Cataloging in Publication Data

Steiner, Rudolf, 1861-1925.
  Cosmic memory.

  (Harper's library of spiritual wisdom)
  Translation of: Aus der Akasha-Chronik.
  Includes index.
  1. Atlantis. 2. Lemuria. 3. Anthroposophy.
I. Title.   II. Series.
BP595.S83663 1981   299'.935   81-47422
ISBN 0-06-067572-1   AACR2

---

81 82 83 84 85 10 9 8 7 6 5 4 3 2 1

THIS VOLUME IS ONE OF THE MAJOR WRITINGS OF
RUDOLF STEINER, ORIGINALLY PUBLISHED IN COM-
MEMORATION OF THE ONE HUNDREDTH ANNIVERSARY
OF HIS BIRTH.

1861-1961

CONTENTS

(7)

Contents

Introduction **RUDOLF STEINER**

THE MAN AND HIS WORK

Rudolf steiner is one of those figures who appear at critical moments in human history, and whose contribution places them in the vanguard of the progress of mankind.

Born in Austria in 1861, educated at the *Technische Hochschule* in Vienna, where he specialized in the study of mathematics and science, Steiner received recognition as a scholar when he was invited to edit the well-known Kürschner edition of the natural scientific writings of Goethe. Already in 1886 at the age of twenty-five, he had shown his comprehensive grasp of the deeper implications of Goethe's way of thinking by writing his *Grundlinien einer Erkenntnistheorie der Goetheschen Weltanschauung* (Theory of Knowledge Implicit in Goethe's Conception of the World). Four years later he was called to join the group of eminent scholars in residence at Weimar, where he worked with them at the Goethe-Schiller Archives for some years. A further result of these activities was the writing of his *Goethes Weltanschauung* (Goethe's Con-

(9)

ception of the World) which, together with his introduc-
tions and commentary on Goethe's scientific writings, established Steiner as one of the outstanding exponents of
Goethe's methodology.

In these years Steiner came into the circle of those
around the aged Nietzsche. Out of the profound impression which this experience made upon him, he wrote his
Friedrich Nietzsche, Ein Kämpfer gegen seine Zeit (Friedrich Nietzsche, a Fighter Against his Time) , published in
1895. This work evaluates the achievements of the great
philosopher against the background of his tragic life-
experience on the one hand, and the spirit of the nineteenth century on the other.

In 1891 Steiner received his Ph.D. at the University of
Rostock. His thesis dealt with the scientific teaching of
Fichte, and is further evidence of Steiner's ability to eval-
uate the work of men whose influence has gone far to
shape the thinking of the modern world. In somewhat enlarged form, this thesis appeared under the title, *Wahrheit
und Wissenschaft* (Truth and Science) , as the preface
to Steiner's chief philosophical work, *Die Philosophie der
Freiheit*, 1894. Later he suggested *The Philosophy of
Spiritual Activity* as the title of the English translation
of this book.

At about this time Steiner began his work as a lec-
turer. This activity was eventually to occupy the major
portion of his time and was to take him on repeated
lecture tours throughout Western Europe. These jour-
neys extended from Norway, Sweden and Finland in the
north to Italy and Sicily in the South, and included sev-

eral visits to the British Isles. From about the turn of the century to his death in 1925, Steiner gave well over 6,000 lectures before audiences of most diverse backgrounds and from every walk of life.

First in Vienna, later in Weimar and Berlin, Steiner wrote for various periodicals and for the daily press. For nearly twenty years, observations on current affairs, reviews of books and plays, along with comment on scientific and philosophical developments flowed from his pen. Finally, upon completion of his work at Weimar, Steiner moved to Berlin in 1897 to assume the editorship of *Das Magazin für Litteratur,* a well-known literary periodical which had been founded by Joseph Lehmann in 1832, the year of Goethe's death.

Steiner's written works, which eventually included over fifty titles, together with his extensive lecturing activity, brought him into contact with increasing numbers of people in many countries. The sheer physical and mental vigor required to carry on a life of such broad, constant activity would alone be sufficient to mark him as one of the most creatively productive men of our time.

The philosophical outlook of Rudolf Steiner embraces such fundamental questions as the being of man, the nature and purpose of freedom, the meaning of evolution, the relation of man to nature, the life after death and before birth. On these and similar subjects, Steiner had unexpectedly new, inspiring and thought-provoking things to say. Through a study of his writings one can come to a clear, reasonable, comprehensive understanding of the human being and his place in the universe.

It is noteworthy that in all his years of work, Steiner made no appeal to emotionalism or sectarianism in his readers or hearers. His scrupulous regard and deep respect for the freedom of every man shines through everything he produced. The slightest compulsion or persuasion he considered an affront to the dignity and ability of the human being. Therefore, he confined himself to objective statements in his writing and speaking, leaving his readers and hearers entirely free to reject or accept his words.

Rudolf Steiner repeatedly emphasized that it is not educational background alone, but the healthy, sound judgment and good will of each individual that enables the latter to comprehend what he has to say. While men and women eminent in cultural, social, political and scientific life have been and are among those who have studied and have found value in Steiner's work, experience has shown repeatedly that his ideas can be grasped by the simplest people. His ability to reach, without exception, all who come to meet his ideas with the *willingness* to understand, is another example of the well-known hallmark of genius.

The ideas of Rudolf Steiner address themselves to the humanity in men and women of every race and of every religious and philosophical point of view, and include them. However, it should be observed that for Steiner the decisive event in world development and the meaning of the historical process is centered in the life and activity of the Christ. Thus, his point of view is essentially Christian, but not in a limited or doctrinal sense. The ideas expressed in his *Das Christentum als mystische Tat-sache und die Mysterien des Altertums* (Christianity as Mystical Fact and the Mysteries of Antiquity) , 1902, and in other works, especially his cycles of lectures on the Gospels (1908-1912) , have brought to many a totally new relationship to Christianity, sufficiently broad to include men of every religious background in full tolerance, yet more deeply grounded in basic reality than are many of the creeds current today.

From his student days, Steiner had been occupied with the education of children. Through his own experience as tutor in Vienna and later as instructor in a school for working men and women in Berlin, he had ample opportunity to gain first-hand experience in dealing with the needs and interests of young people. In his Berlin teaching work he saw how closely related are the problems of education and of social life. Some of the fundamental starting-points for an educational praxis suited to the needs of children and young people today, Steiner set forth in a small work titled *Die Erziehung des Kindes vom Gesichtspunkte der Geisteswissenschaft* (The Education of the Child in the Light of the Science of the Spirit) , published in 1907.

Just forty years ago, in response to an invitation arising from the need of the time and from some of the ideas expressed in the essay mentioned above, Rudolf Steiner inaugurated a system of education of children and young people based upon factors inherent in the nature of the growing child, the learning process, and the requirements of modern life. He himself outlined the curricu-

lum, selected the faculty, and, despite constant demands
for his assistance in many other directions, he carefully
supervised the initial years of activity of the first Rudolf
Steiner Schools in Germany, Switzerland and England.
The story of the successful development of the educational movement over the past forty years cannot be
told here. However, from the opening of the first Rudolf
Steiner School, the Waldorf School in Stuttgart, Germany, to the present time, the success of Rudolf Steiner
Education (sometimes referred to as Waldorf Education) has proven the correctness of Steiner's concept of
the way in which to prepare the child for his eventual
adult role in and his contribution to modern society.
Today there are some seventy Rudolf Steiner Schools in
existence in seventeen countries of the world, including
the United States, Canada, Mexico, and South America.

In 1913, at Dornach near Basel, Switzerland, Rudolf
Steiner laid the foundation of the Goetheanum, a unique
building erected in consonance with his design and under
his personal supervision. Intended as the building in
which Steiner's four dramas would be performed, the
Goetheanum also became the center of the Anthroposophical Society which had been founded by students
of Rudolf Steiner in 1912. The original building was
destroyed by fire in 1922, and subsequently was replaced
by the present Goetheanum, also created from designs
prepared by Rudolf Steiner.

Today the Goetheanum is the world headquarters of
the General Anthroposophical Society, which was founded

at Dornach at Christmas, 1923, with Rudolf Steiner as President. Audiences of many thousands come there each year to attend performances of Steiner's dramas, of Goethe's *Faust* (Parts I and II in their entirety), and of plays by other authors, presented on the Goetheanum stage, one of the finest in Europe. Eurythmy performances, musical events, conferences and lectures on many subjects, as well as courses of study in various fields attract people to the Goetheanum from many countries of the world, including the United States.

Among activities springing from the work of Rudolf Steiner are Bio-Dynamic Farming and Gardening, which aims at improved nutrition resulting from methods of agriculture outlined by him; the art of Eurythmy, created and described by him as "visible speech and visible song"; the work of the Clinical and Therapeutical Institute at Arlesheim, Switzerland, with related institutions in other countries, where for the past thirty years the indications given by Rudolf Steiner in the fields of Medicine and Pharmacology have been applied; the Homes for Children in need of special care, which exist in many countries for the treatment of mentally retarded children along lines developed under Steiner's direction; the further development of Steiner's indications of new directions of work in such fields as Mathematics, Physics, Painting, Sculpture, Music Therapy, Drama, Speech Formation, Astronomy, Economics, Psychology, and so on. Indeed, one cannot but wonder at the breadth, the scope of the benefits which have resulted from the work of this one man!

A full evaluation of what Rudolf Steiner accomplished for the good of mankind in so many directions can come about only when one comprehends the ideas which motivated him. He expressed these in his writings, of which the present volume is one. Taken together, these written works comprise the body of knowledge to which Steiner gave the name, the science of the spirit, or Anthroposophy. On page 249 of this book he writes of the benefits of this science of the spirit:

"When correctly understood, the truths of the science of the spirit will give man a true foundation for his life, will let him recognize his value, his dignity, and his essence, and will give him the highest zest for living. For these truths enlighten him about his connection with the world around him; they show him his highest goals, his true destiny. And they do this in a way which corresponds to the demands of the present, so that he need not remain caught in the contradiction between belief and knowledge."

*

Many of the thoughts expressed in this book may at first appear startling, even fantastic in their implications. Yet when the prospect of space travel, as well as modern developments in technology, psychology, medicine and philosophy challenge our entire understanding of life and the nature of the living, strangeness as such should be no valid reason for the serious reader to turn away from a book of this kind. For example, while the word "occult" or "supersensible" may have undesirable conno-

tations for many, current developments are fast bringing
re-examination of knowledge previously shunned by con-
ventional research. The challenge of the atomic age has
made serious re-evaluation of all knowledge imperative,
and it is recognized that no single area of that knowledge
can be left out of consideration.

Steiner himself anticipated the reader's initial difficulties with this book, as he indicates on page 112: "The
reader is requested to bear with much that is dark and
difficult to comprehend, and to struggle toward an understanding, just as the writer has struggled toward a generally understandable manner of presentation. Many a
difficulty in reading will be rewarded when one looks
upon the deep mysteries, the important human enigmas
which are indicated."

On the other hand, a further problem arises as a result
of Steiner's conviction regarding the purpose for which
a book dealing with the science of the spirit is designed.
This involves the *form* of the book as against its content. Steiner stressed repeatedly that a book on the
science of the spirit does not exist only for the purpose
of conveying information to the reader. With painstaking
effort, he elaborated his books in such a manner that
while the reader receives certain information from the
pages, he also experiences a kind of *awakening* of spiritual
life within himself. Steiner describes this awakening as
". . . an experiencing with inner shocks, tensions and
resolutions." In his autobiography he speaks of his striving to bring about such an awakening in the readers of
his books: "I know that with every page my inner battle

has been to reach the utmost possible in this direction. In the matter of style, I do not so describe that my subjective feelings can be detected in the sentences. In writing I subdue to a dry mathematical style what has come out of warm and profound feeling. But only such a style can be an awakener, for the reader must cause warmth and feeling to awaken in himself. He cannot simply allow these to flow into him from the one setting forth the truth, while he remains passively composed." (*The Course of My Life*, p. 330)

In the present translation, therefore, careful effort has been made to preserve as much as possible such external form details as sentence and paragraph arrangement, italics, and even some of the more characteristic punctuation of the original, regardless of currently accepted English usage.

*

The essays contained in this book occupy a significant place in the life-work of Rudolf Steiner. They are his first written expression of a cosmology resulting from that spiritual perception which he described as "a fully conscious standing-within the spiritual world." In his autobiography he refers to the early years of the present century as the time when, "Out of the experience of the spiritual world in general developed specific details of knowledge." (*Op. cit.* pp. 326, 328.) Steiner has stated that from his early childhood he knew the reality of the spiritual world because he could experience this spiritual world directly. However, only after nearly forty years was it possible for him to transmit to others concrete, detailed information regarding this spiritual world.

As they appear in the present essays, these "specific details" touch upon processes and events of extraordinary sweep and magnitude. They include essential elements of man's prehistory and early history, and shed light upon the evolutionary development of our earth. Published now for the first time in America, just a century after Darwin's *Origin of the Species* began its transformation of man's view of himself and of his environment, these essays clarify and complement the pioneer work of the great English scientist.

Rudolf Steiner shows that the insoluble link between man and cosmos is the fundamental basis of evolution. As man has participated in the development of the world we know today, so his achievements are directly connected with the ultimate destiny of the universe. In his hands rests the freedom to shape the future course of creation. Knowledge of his exalted origins and of the path he followed in forfeiting divine direction for the attainment of his present self-dependent freedom, are indispensable if man is to evolve a future worthy of a responsible human being.

This book appears now because of its particular significance at a moment when imperative and grave decisions are being made in the interests of the future of mankind.

PAUL MARSHALL ALLEN

Englewood, New Jersey
June 1959

PREFACE BY THE EDITOR
OF THE GERMAN EDITION (1939)

THESE ESSAYS of Dr. Rudolf Steiner which first appeared in 1904 are now published in book form after thirty-five years. They were written for the periodical *Lucifer Gnosis,* which appeared at first as a monthly and then at longer intervals. This explains the occasional repetition of what has been said previously. But, after all, repetitions are especially useful in the study of the science of the spirit. However, some may find it confusing that beside the new terminology coined for the Occident one is also mentioned which has been taken from oriental esoterica. The latter had become popular in Europe around the turn of the century through the literature of the Theosophical Society. The exotic names had stayed in people's memories, but the finer nuances which the Oriental associates with them remained closed to the European. The adaptation of our language, which is fitted for sensory perception, to a more delicate spiritual conceptualization and to a concrete picturing of even the extra-sensory was something at which Dr. Steiner worked unceasingly. In the description of the activity of the Hierarchies he uses the Christian terminology customary for this purpose.

What is here presented in form of a brief survey, finds its continuation in the books *Theosophie* and *Geheimwissenschaft im Umriss*.

The periodical *Lucifer Gnosis* could not be continued because of the excessive demands made by lecturing activities and other occupations. Apart from the results of spiritual scientific research, it contained many essays in which Dr. Steiner comes to grips with the scientific thinking of the present. Since writings like these concerning the Akasha Chronicle cannot fail to appear as wild phantasy to most unprepared readers of today, two essays from this periodical which touch upon the epistemological problems of the present, precede and follow them. The sober logic of these two essays should furnish proof that the investigator of supersensible worlds is also able to survey problems of the present in a calm and objective manner.

The periodical was also devoted to the answering of questions posed by its readers. From this section we include some points relating to Atlantean humanity and to mystery science. However, the one who wishes to obtain a clear idea of the manner in which a reading of the Akasha Chronicle becomes possible, must devote himself intensively to the study of Anthroposophy.

Apart from the above-mentioned books we indicate for those who are advanced in the study of the science of the spirit, the esoteric reflections on *Okkultes Lesen und Okkultes Hören* (Occult Reading and Occult Hearing), and the third volume of the series *Geistige Wesen und Ihre*

Wirkungen (Spiritual Beings and Their Effects) which has just appeared and should be of special interest today: *Geschichtliche Notwendigkeit und Freiheit, Schicksalseinwirkungen aus der Welt der Toten* (Historical Necessity and Freedom, Fateful Influences from the World of the Dead).

MARIE STEINER (1867-1948)

Cosmic Memory

PREHISTORY OF EARTH AND MAN

i CONTEMPORARY CIVILIZATION IN THE MIRROR OF THE SCIENCE OF THE SPIRIT (1904)

THE OBSERVER of the course of scientific development in the last decades cannot doubt that a great revolution is in preparation. Today when a scientist talks about the so-called enigmas of existence, it sounds quite different than it did a short time ago.

Around the middle of the nineteenth century some of the most daring spirits saw in scientific materialism the only creed possible to one familiar with the then recent results of research. The blunt saying of that time has become famous: "Thoughts stand in about the same relationship to the brain as gall to the liver." This was stated by Karl Vogt, who in his *Köhlerglauben und Wissenschaft* (Blind Faith and Science) and in other writings, declared everything to be superannuated which did not make spiritual activity and the life of the soul proceed from the mechanism of the nervous system and of the brain in the same manner in which the physicist explains that the movement of the hands proceeds from the mecha-

nism of the clock. That was the time when Ludwig
Buechner's *Kraft und Stoff* (Force and Matter) became
a sort of gospel among wide circles of the educated. One
may well say that excellent, independently thinking minds
came to such convictions because of the powerful impression made by the successes of science in those times. A
short time before, the microscope had shown the syn-
thesis of living beings out of their smallest parts, the cells.
Geology, the science of the formation of the earth, had
come to the point of explaining the development of the
planets in terms of the same laws which still operate today.
Darwinism promised to explain the origin of man in a
completely natural way and began its victorious course
through the educated world so auspiciously that for many
it seemed to dispose of all "old belief." A short time ago,
all this became quite different. It is true that stragglers
who adhere to these opinions can still be found in men
like Ladenburg at the Congress of Scientists in 1903, who
proclaim the materialistic gospel; but against them stand
others who have arrived at a quite different way of speaking through more mature reflection on scientific questions. A work has just appeared which bears the title,
Naturwissenschaft und Weltanschauung (Science and
World Conception) . Its author is Max Verworn, a physiol-
ogist of the school of Haeckel. In this work one can read
the following: "Indeed, even if we possessed the most
complete knowledge of the physiological events in the
cells and fibers of the cerebral cortex with which psychic
events are connected, even if we could look into the
mechanism of the brain as we look into the works of a

clock, we would never find anything but moving atoms. No human being could see or otherwise perceive through his senses how sensations and ideas arise in this mechanism. The results which the materialistic conception has obtained in its attempt to trace mental processes back to the movements of atoms illustrates its efficiency very clearly. As long as the materialistic conception has existed, it has not explained the simplest sensation by movements of atoms. Thus it has been and thus it will be in future. How could it be conceivable that things which are not perceptible by the senses, such as the psychic processes, could ever be explained by a mere splitting up of large bodies into their smallest parts? The atom is still a body after all, and no movement of atoms is ever capable of bridging the gulf between the material world and the psyche. However fruitful the materialistic point of view has been as a scientific working hypothesis, however fruitful it will doubtless remain in this sense in the future—I point only to the successes of structural chemistry—just as useless is it as the basis for a world conception. Here it shows itself to be too narrow. Philosophical materialism has finished playing its historical role. This attempt at a scientific world conception has failed for ever." Thus, at the beginning of the twentieth century, a scientist speaks about the conception which around the middle of the nineteenth was proclaimed as a new gospel demanded by the advances of science.

It is especially the 'fifties, the 'sixties, and the 'seventies of the nineteenth century which may be designated as the years of the high tide of materialism. The explanation of mental and spiritual phenomena on the basis of purely mechanical processes exercised a really fascinating influence at that time. The materialists could tell themselves that they had won a victory over the adherents of a spiritual world conception. Those also who had not started from scientific studies joined their ranks. While Buechner, Vogt, Moleschott and others still built on purely scientific premises, in his *Alten und neuen Glauben* (Old and New Belief, 1872), David Friedrich Strauss attempted to obtain bases for the new creed from his theological and philosophical ideas. Decades before he had already intervened in the intellectual life with his *Leben Jesu* (Life of Jesus) in a manner which caused a sensation. He seemed to be equipped with the full theological and philosophical culture of his time. He now said boldly that the materialistic explanation of the phenomena of the universe, including man, had to form the basis for a new gospel, for a new moral comprehension and formation of existence. The descent of man from purely animal ancestors seemed about to become a new dogma, and in the eyes of scientific philosophers, all adherence to spiritual-soul origin of our race amounted to an antiquated superstition from the infancy of mankind, with which one did not have to disturb oneself further.

The historians of culture came to the aid of those who built on the new science. The customs and ideas of savage tribes were made the object of study. The remains of primitive cultures, which are dug out of the ground like the bones of prehistoric animals and the impressions of extinct plants were to bear witness to the fact that at his first appearance on earth man was distinguished only in degree from the higher animals, and that mentally and spiritually he had risen to his present eminence from the level of animalism pure and simple. A time had come when everything in this materialistic edifice seemed to be right. Under a kind of coercion which the ideas of the time exercised on them, men thought as a faithful materialist has written: "The assiduous study of science has brought me to the point where I accept everything calmly, bear the inevitable patiently, and for the rest help in the work of gradually reducing the misery of mankind. The fantastic consolations which a credulous mind seeks in marvelous formulas I can renounce all the more easily since my imagination receives the most beautiful stimulation through literature and art. When I follow the plot of a great drama or, under the guidance of scientists, make a journey to other stars, an excursion through prehistoric landscapes, when I admire the majesty of nature on mountain peaks or venerate the art of man in tones and colors, do I not then have enough of the elevating? Do I then still need something which contradicts my reason? The fear of death, which torments so many of the pious, is completely unknown to me. I know that I no more survive after my body decays than I lived before my birth. The agonies of purgatory and of hell do not exist for me. I return to the boundless realm of Nature, who embraces all her children lovingly. My life was not in vain. I have made good use of the strength which I possessed. I depart from earth in the firm belief that everything will become better and more beautiful." *Vom*

*Glauben zum Wissen. Ein lehrreicher Entwickelungsgang
getreu nach dem Leben geschildert von Kuno Freidank.*
(On the Belief in Knowledge. An Instructive Course of
Development Described in a Manner Faithfully True to
Life by Kuno Freidank.) Many people who are still subject to the compulsive ideas which acted upon the representatives of the materialistic world conception in the
time mentioned above, also think in this manner today.

Those however who tried to maintain themselves on
the heights of scientific thought have come to other ideas.
The first reply to scientific materialism, made by an eminent scientist at the Congress of Scientists in Leipzig
(1876), has become famous. Du Bois-Reymond at that
time made his *"Ignorabimus* speech." He tried to demonstrate that this scientific materialism could in fact do
nothing but ascertain the movements of the smallest material particles, and he demanded that it should be satisfied with doing this. But he emphasized at the same time
that in doing this it contributes absolutely nothing to
an explanation of mental and spiritual processes. One
may take whatever attitude one pleases toward these
statements of Du Bois-Reymond, but this much is clear:
they represented a rejection of the materialistic interpretation of the world. They showed how as a scientist one
could lose confidence in this interpretation.

The materialistic interpretation of the world had thereby
entered the stage where it declared itself to be unassum-
ing as far as the life of the soul is concerned. It admitted
its "ignorance" (agnosticism). It is true that it declared
its intention of remaining "scientific" and of not having

recourse to other sources of knowledge, but on the other hand it did not want to ascend with its means to a higher world-conception. In recent times Raoul Francé, a scientist, has shown in comprehensive fashion the inadequacy of scientific results for a higher world-conception. This is an undertaking to which we would like to refer again on another occasion.

The facts now steadily increased which showed the impossibility of the attempt to build up a science of the soul on the investigation of material phenomena. Science was forced to study certain "abnormal" phenomena of the life of the soul like hypnotism, suggestion, somnambulism. It became apparent that in the face of these phenomena a materialistic view is completely inadequate for a truly thinking person. The facts with which one became acquainted were not new. They were phenomena which had already been studied in earlier times and up to the beginning of the nineteenth century, but which in the time of the materialistic flood had simply been put aside as inconvenient.

To this was added something else. It became more and more apparent on how weak a basis the scientists had built, even as far as their explanations of the origin of animal species and consequently of man were concerned. For a while, the ideas of "adaptation" and of the "struggle for existence" had exercised an attraction in the explanation of the origin of species. One learned to understand that in following them one had followed mirages. A school was formed under the leadership of Weismann which denied that characteristics which an organism had *acquired*

through adaptation to the environment could be transmitted by inheritance, and that in this way a *transformation*
of organisms could occur. One therefore ascribed everything to the "struggle for existence" and spoke of an "om-
nipotence of natural selection." A stark contrast to this
view was presented by those who, relying on unquestion-
able facts, declared that a "struggle for existence" had been
spoken of in cases where it did not even exist. They wanted
to demonstrate that nothing could be explained by it.
They spoke of an "impotence of natural selection." Moreover, in the last years de Vries was able to show experi-
mentally that changes of one life-time into another can occur *by leaps,* mutation. With this, what was regarded as a
firm article of faith by the Darwinists, namely that animal
and plant forms change only gradually, was shaken. More
and more the ground on which one had built for decades
simply disappeared beneath one's feet. Even earlier, thinking scientists had realized that they had to abandon this
ground; thus W. H. Rolph, who died young, in 1884
declared in his book, *Biologische Probleme, zugleich als
Versuch zur Entwicklung einer rationellen Ethik* (Bio-
logical Problems, with an Attempt at the Development
of Rational Ethics) : "Only through the introduction of
insatiability does the Darwinian principle of the struggle for life become acceptable. Because it is only then
that we have an explanation for the fact that wherever
it can, a creature acquires more than it needs for main-
taining the *status quo,* that it grows to excess where the
occasion for this is given. . . . While for the Darwinists
there is no struggle for existence wherever the existence

of a creature is not threatened, for me the struggle is an omnipresent one. It is primarily a struggle for life, a struggle for the increase of life, not a struggle for existence."

It is only natural that in view of these facts the judicious confess to themselves: "The materialistic universe of thought is not fit for the construction of a world-conception. If we base ourselves on it, we cannot say anything about mental and spiritual phenomena." Today there are already numerous scientists who seek to erect a structure of the world for themselves, based on quite different ideas. One need only recall the work of the botanist, Reincke, *Die Welt als Tat* (The World as Deed). However, it becomes apparent that such scientists have not been trained with impunity amidst purely materialistic ideas. What they utter from their new idealistic standpoint is inadequate, can satisfy them for a while, but not those who look more deeply into the enigmas of the world. Such scientists cannot bring themselves to approach those methods which proceed from a real contemplation of the mind and the soul. They have the greatest fear of "mysticism," of "gnosis" or "theosophy." This appears clearly, for example, in the work of Verworn quoted above. He says: "There is a ferment in science. Things which seemed clear and transparent to everybody have become cloudy today. Long-tested symbols and ideas, with which everyone dealt and worked at every step without hesitation a short time ago, have begun to totter and are looked upon with suspicion. Fundamental concepts, such as those of matter, appear to have been shaken, and the

firmest ground is beginning to sway under the scientist's
steps. Certain problems alone stand with rocklike firm-
ness, problems on which until now all attempts, all efforts of science have been shattered. In the face of this
knowledge one who is despondent resignedly throws himself into the arms of mysticism, which has always been
the last refuge when the tormented intellect could see
no way out."The sensible man looks for new symbols
and attempts to create new bases on which he can build
further." One can see that because of his habits of conceptualization the scientific thinker of today is not in a
position to think of "mysticism" otherwise than as im-
plying intellectual confusion and vagueness. What concepts of the life of the soul does such a thinker not reach!
At the end of the work referred to above, we read: "Prehistoric man formed the idea of a separation of body and
soul in face of death. The soul separated itself from the
body and led an independent existence. It found no rest
and returned as a ghost unless it was banned by sepulchral
ceremonies. Man was terrorized by fear and superstition.
The remains of these ideas have come down to our time.
The fear of death, that is, of what is to come after, is
widespread today. How differently does all this appear
from the standpoint of psychomonism! Since the psychic
experiences of the individual only take place when certain regular connections exist, they cease when these
connections are in any way disturbed, as happens number-
less times in the course of a day. With the bodily changes
at death, these connections stop entirely. Thus, no sensation and conception, no thought and no feeling of the

individual can remain. The *individual* soul is dead. Nevertheless the sensations and thoughts and feelings continue to live. They live beyond the transitory individual in other individuals, wherever the same complexes of conditions exist. They are transmitted from individual to individual, from generation to generation, from people to people. They weave at the eternal loom of the soul. They work at the history of the human spirit. "Thus we all survive after death as links in the great interconnected chain of spiritual development." But is that something different from the survival of the wave in others which it has caused, itself meanwhile disappearing? Does one really survive when one continues to exist only in one's effects? Does one not have such a survival in common with all phenomena, even those of physical nature? One can see that the materialistic world conception had to undermine its own foundations. As yet it cannot lay new ones. Only a true understanding of mysticism, theosophy, and gnosis will enable it to do so. The chemist Osterwald spoke several years ago at the Congress of Scientists at Luebeck of the "overcoming of materialism," and for this purpose founded a new periodical dealing with the philosophy of nature. Science is ready to receive the fruits of a higher world-conception. All resistance will avail it nothing; it will have to take into account the needs of the longing human soul.

PREFACE

By MEANS OF ordinary history man can learn only a small part of what humanity experienced in prehistory. Historical documents shed light on but a few millennia. What archaeology, paleontology, and geology can teach us is very limited. Furthermore, everything built on external evidence is unreliable. One need only consider how the picture of an event or people, not so very remote from us, has changed when new historical evidence has been discovered. One need but compare the descriptions of one and the same thing as given by different historians, and he will soon realize on what uncertain ground he stands in these matters. Everything belonging to the external world of the senses is subject to time. In addition, time destroys what has originated in time. On the other hand, external history is dependent on what has been preserved in time. Nobody can say that the essential has been preserved, if he remains content with external evidence.

Everything which comes into being in time has its origin in the eternal. But the eternal is not accessible to

sensory perception. Nevertheless, the ways to the perception of the eternal are open for man. He can develop
forces dormant in him so that he can recognize the eternal.
In the essays, *Wie erlangt man Erkenntnisse der höheren
Welten?* (How Does One Attain Knowledge of Higher
Worlds?), which appear in this periodical*, this development is referred to. These present essays will also show
that at a certain high level of his cognitive power, man
can penetrate to the eternal origins of the things which
vanish with time. A man broadens his power of cognition
in this way if he is no longer limited to external evidence where knowledge of the past is concerned. Then
he can *see* in events what is not perceptible to the senses,
that part which time cannot destroy. He penetrates from
transitory to non-transitory history. It is a fact that this
history is written in other characters than is ordinary history. In gnosis and in theosophy it is called the "Akasha
Chronicle." Only a faint conception of this chronicle can
be given in our language. For our language corresponds
to the world of the senses. That which is described by
our language at once receives the character of this sense
world. To the uninitiated, who cannot yet convince himself of the reality of a separate spiritual world through
his own experience, the initiate easily appears to be a
visionary, if not something worse.

The one who has acquired the ability to perceive in
the spiritual world comes to know past events in their
eternal character. They do not stand before him like the
dead testimony of history, but appear in full *life*. In a

---

* These essays were published in book form, Berlin, 1909.

certain sense, what has happened takes place before him.

Those initiated into the reading of such a living script can look back into a much more remote past than is represented by external history; and—on the basis of direct spiritual perception—they can also describe much more dependably the things of which history tells. In order to avoid possible misunderstanding, it should be said that spiritual perception is not infallible. This perception also can err, can see in an inexact, oblique, wrong manner. No man is free from error in this field, no matter how high he stands. Therefore one should not object when communications emanating from such spiritual sources do not always entirely correspond. But the dependability of observation is much greater here than in the external world of the senses. What various initiates can relate about history and prehistory will be in *essential* agreement. Such a history and prehistory does in fact exist in all mystery schools. Here for millennia the agreement has been so complete that the conformity existing among external historians of even a single century cannot be compared with it. The initiates describe *essentially* the same things at all times and in all places.

Following this introduction, several chapters from the Akasha Chronicle will be given. First, those events will be described which took place when the so-called *Atlantean* Continent still existed between America and Europe. This part of our earth's surface was once land. Today this forms the floor of the Atlantic Ocean. Plato tells of the last remnant of this land, the island Poseidon, which lay westward of Europe and Africa. In *The Story of*

Atlantis and Lost Lemuria, by W. Scott-Elliott, the reader
can find that the floor of the Atlantic Ocean was once a
continent, that for about a million years it was the scene
of a civilization which, to be sure, was quite different
from our modern ones, and the fact that the last remnants
of this continent sank in the tenth millennium B.C. In
this present book the intention is to give information
which will supplement what is said by Scott-Elliott.
While he describes more the outer, the external events
among our Atlantean ancestors, the aim here is to record
some details concerning their spiritual character and the
inner nature of the conditions under which they lived.
Therefore the reader must go back in imagination to a
period which lies almost ten thousand years behind us,
and which lasted for many millennia. What is described
here however, did not take place only on the continent
now covered by the waters of the Atlantic Ocean, but
also in the neighboring regions of what today is Asia,
Africa, Europe, and America. What took place in these
regions later, developed from this earlier civilization.

Today I am still obliged to remain silent about the
sources of the information given here. One who knows
anything at all about such sources will understand why
this has to be so. But events can occur which will make
a breaking of this silence possible very soon. How much
of the knowledge hidden within the theosophical movement may gradually be communicated, depends entirely
on the attitude of our contemporaries.

Now follows the first of the writings which can be
given here.

iii OUR ATLANTEAN ANCESTORS

Our Atlantean ancestors differed more from present-day man than he would imagine whose knowledge is confined wholly to the world of the senses. This difference extended not only to the external appearance but also to spiritual faculties. Their knowledge, their technical arts, indeed their entire civilization differed from what can be observed today. If we go back to the first periods of Atlantean humanity we find a mental capacity quite different from ours. Logical reason, the power of arithmetical combining, on which everything rests that is produced today, were totally absent among the first Atlanteans. On the other hand, they had a highly developed *memory*. This memory was one of their most prominent mental faculties. For example, the Atlantean did not calculate as we do, by learning certain rules which he then applied. A "multiplication table" was something totally unknown in Atlantean times. Nobody impressed upon his intellect that three times four is twelve. In the event that he had to perform such a calculation he could manage because he remembered identical or similar situations. He *remembered* how it had been on previous occasions. One need only realize that each time a new faculty develops in an organism, an old faculty loses power and acuteness. The man of today is superior to the Atlantean in logical reasoning, in the ability to combine. On the other hand, memory has deteriorated. Nowadays man thinks in concepts; the Atlantean thought in images. When an image appeared in his soul he remembered a great many similar images which he had already experienced. He directed his judgment accordingly. For this reason all teaching at that time was different from what it became later. It was not calculated to furnish the child with rules, to sharpen his reason. Instead, life was presented to him in vivid images, so that later he could remember as much as possible when he had to act under particular conditions. When the child had grown and had gone out into life, for everything he had to do he could remember something similar which had been presented to him in the course of his education. He could manage best when the new situation was similar to one he had already seen. Under totally new conditions the Atlantean had to rely on experiment, while in this respect much has been spared modern man due to the fact that he is equipped with rules. He can easily apply these in those situations which are new to him. The Atlantean system of education gave a uniformity to all of life. For long periods things were done again and again in the same way. The faithful memory did not allow anything to develop which was even remotely similar to the rapidity of our present-day progress. One did what one had always "seen" before. One did not invent; one remembered. He

was not an authority who had learned much, but rather
he who had experienced much and therefore could remember much. In the Atlantean period it would have
been impossible for someone to decide an important
matter before reaching a certain age. One had confidence
only in a person who could look back upon long experience.

What has been said here was not true of the initiates
and their schools. For *they* are in advance of the stage
of development of their period. For admission into such
schools, the decisive factor is not age, but whether in his
previous incarnations the applicant has acquired the faculties for receiving higher wisdom. The confidence placed
in the initiates and their representatives during the Atlantean period was not based on the richness of their
personal experience, but rather on the *antiquity* of their
wisdom. In the case of the initiate, personality ceases to
have any importance. He is totally in the service of *eternal*
wisdom. Therefore the characteristic features of a particular period do not apply to him.

While the power to think logically was absent among
the Atlanteans (especially the earlier ones) , in their highly
developed memory they possessed something which gave
a special character to everything they did. But with the
nature of one human power others are always connected.
Memory is closer to the deeper natural basis of man than
reason, and in connection with it other powers were
developed which were still closer to those of subordinate
natural beings than are contemporary human powers.
Thus the Atlanteans could control what one calls the *life*

force. As today one extracts the energy of heat from coal and transforms it into motive power for our means of locomotion, the Atlanteans knew how to put the germinal energy of organisms into the service of their technology. One can form an idea of this from the following. Think of a kernel of seed-grain. In this an energy lies dormant. This energy causes the stalk to sprout from the kernel. Nature can awaken this energy which reposes in the seed. Modern man cannot do it at will. He must bury the seed in the ground and leave the awakening to the forces of nature. The Atlantean could do something else. He knew how one can change the energy of a pile of grain into technical power, just as modern man can change the heat energy of a pile of coal into such power. Plants were cultivated in the Atlantean period not merely for use as foodstuffs but also in order to make the energies dormant in them available to commerce and industry. Just as we have mechanisms for transforming the energy dormant in coal into energy of motion in our locomotives, so the Atlanteans had mechanisms in which they—so to speak—burned plant seeds, and in which the life force was transformed into technically utilizable power. The vehicles of the Atlanteans, which floated a short distance above the ground, were moved in this way. These vehicles travelled at a height lower than that of the mountain ranges of the Atlantean period, and they had steering mechanisms by the aid of which they could rise above these mountain ranges.

One must imagine that with the passage of time all conditions on our earth have changed very much. Today, the above-mentioned vehicles of the Atlanteans would be totally useless. Their usefulness depended on the fact that then the cover of air which envelops the earth was *much denser* than at present. Whether in face of current scientific beliefs one can easily imagine such greater density of air, must not occupy us here. Because of their very nature, science and logical thinking can never decide what is possible or impossible. Their only function is to explain what has been ascertained by experience and observation. The above-mentioned density of air is as certain for occult experience as any fact of today given by the senses can be.

Note →

Equally certain however is the fact, perhaps even more inexplicable for contemporary physics and chemistry, that at that time the *water* on the whole earth was much *thinner* than today. Because of this thinness the water could be directed by the germinal energy used by the Atlanteans into technical services which today are impossible. As a result of the increased density of the water, it has become impossible to move and to direct it in such ingenious ways as once were possible. From this it must be sufficiently clear that the civilization of the Atlantean period was radically different from ours. It will also be understood that the physical nature of an Atlantean was quite different from that of a contemporary man. The Atlantean took into himself water which could be used by the life force inherent in his own body in a manner quite different from that possible in today's physical body. It was due to this that the Atlantean could consciously employ his physical powers in an entirely different way

Note on water much

from a man of today. He had, so to speak, the means to increase the physical powers in himself when he needed them for what he was doing. In order to have an accurate conception of the Atlanteans one must know that their ideas of fatigue and the depletion of forces were quite different from those of present-day man.

An Atlantean settlement—as must be evident from everything we have described—had a character which in no way resembled that of a modern city. In such a settlement everything was, on the contrary, still in alliance with nature. Only a vaguely similar *picture* is given if one should say that in the first Atlantean periods—about to the middle of the third subrace—a settlement resembled a garden in which the houses were built of trees with artfully intertwined branches. What the work of human hands created at that time grew out of nature. And man himself felt wholly related to nature. Hence his social sense also was quite different from that of today. After all, nature is common to all men. What the Atlantean built up on the basis of nature he considered to be *common property* just as a man of today thinks it only natural to consider as his private property what his ingenuity, his intelligence have created for him.

One familiar with the idea that the Atlanteans were equipped with such spiritual and physical powers as have been described, will also understand that in still earlier times mankind presented a picture which reminds him in only a few particulars of what he is accustomed to see today. Not only men, but also the surrounding nature has changed enormously in the course of time. Plant and

[margin note: note on Atlantean house]

[margin note: Change]

animal forms have become different. All of earthly nature
has been subjected to transformations. Once inhabited
regions of earth have been destroyed; others have come
into existence.

The ancestors of the Atlanteans lived in a region which
has disappeared, the main part of which lay south of
contemporary Asia. In theosophical writings they are
called the Lemurians. After they had passed through various stages of development the greatest part of them declined. These became stunted men, whose descendants
still inhabit certain parts of the earth today as so-called
savage tribes. Only a small part of Lemurian humanity
was capable of further development. From this part the
Atlanteans were formed.

Later, something similar again took place. The greatest part of the Atlantean population declined, and from
a small portion are descended the so-called Aryans who
comprise present-day civilized humanity. According to
the nomenclature of the science of the spirit, the Lemurians, Atlanteans and Aryans are *root races* of mankind. If
one imagines that two such root races preceded the Lemurians and that two will succeed the Aryans in the future,
one obtains a total of *seven*. One always arises from another in the manner just indicated with respect to the
Lemurians, Atlanteans, and Aryans. Each root race has
physical and mental characteristics which are quite different from those of the preceding one. While, for example, the Atlanteans especially developed memory and
everything connected with it, at the present time it is

the task of the Aryans to develop the faculty of thought and all that belongs to it.

In each root race various stages must also be gone through. There are always seven of these. In the beginning of a period identified with a root race, its principal characteristics are in a youthful condition; slowly they attain maturity and finally enter a decline. The population of a root race is thereby divided into seven subraces. But one must not imagine that one subrace immediately disappears when a new one develops. Each one may maintain itself for a long time while others are developing beside it. Thus there are always populations which show different stages of development living beside each other on earth.

The first subrace of the Atlanteans developed from a very advanced part of the Lemurians who had a high evolutionary potential. The faculty of memory appeared only in its rudiments among the Lemurians, and then only in the last period of their development. One must imagine that while a Lemurian could form ideas of what he was experiencing, he could not preserve these ideas. He immediately forgot what he had represented to himself. Nevertheless, that he lived in a certain civilization, that, for example, he had tools, erected buildings and so forth—this he owed not to his *own* powers of conception, but to a mental force in him, which was *instinctive*. However, one must not imagine this to have been the present-day instinct of animals, but one of a different kind.

Theosophical writings call the first subrace of the Atlanteans that of the Rmoahals. The memory of this race
was primarily directed toward vivid sense impressions.
Colors which the eye had seen, sounds which the ear
had heard, had a long after-effect in the soul. This was
expressed in the fact that the Rmoahals developed *feelings* which their Lemurian ancestors did not yet know.
For example, the attachment to what has been experienced in the past is a part of these feelings.

With the development of memory was connected that
of *language*. As long as man did not preserve what was
past, a communication of what had been experienced
could not take place through the medium of language.
Because in the last Lemurian period the first beginnings
of memory appeared, at that time it was also possible for
the faculty of naming what had been seen and heard to
have its inception. Only men who have the faculty of
recollection can make use of a name which has been given
to something. The Atlantean period, therefore, is the one
in which the development of language took place. With
language a bond was established between the human soul
and the things outside man. He produced a speech-word
inside himself, and this speech-word belonged to the
objects of the external world. A new bond is also formed
among men by communications through the medium of
language. It is true that all this existed in a still youthful
form among the Rmoahals, but nevertheless it distinguished them profoundly from their Lemurian fore-
fathers.

The soul powers of these first Atlanteans still possessed something of the forces of nature. These men were more closely related to the beings of nature which surrounded them than were their successors. Their soul powers were more connected with forces of nature than are those of modern man. Thus the speech-word which they produced had something of the power of nature. They not only *named* things, but in their words was a *power* over things and also over their fellow-men. The word of the Rmoahals not only had *meaning*, but also *power*. The magic power of words is something which was far truer for those men than it is for men of today. When a Rmoahals man pronounced a word, this word developed a power similar to that of the object it designated. Because of this, words at that time were curative; they could advance the growth of plants, tame the rage of animals, and perform other similar functions. All this progressively decreased in force among the later subraces of the Atlanteans. One could say that the original fullness of power was gradually lost. The Rmoahals men felt this plenitude of power to be a gift of mighty nature, and their relationship to the latter had a religious character. For them language was something especially sacred. The misuse of certain sounds, which possessed an important power, was an impossibility. Each man felt that such misuse must cause him enormous harm. The good magic of such words would have changed into its opposite; that which would have brought blessings if used properly would bring ruin to the author if used criminally. In a kind of inno-

cence of feeling the Rmoahals ascribed their power not
so much to themselves as to the *divine nature* acting
within them.

This changed among the second subrace, the so-called
Tlavatli peoples. The men of this race began to feel
their own personal value. Ambition, a quality unknown
to the Rmoahals, made itself felt among them. *Memory*
was in a sense transferred to the conception of communal
life. He who could look back upon certain deeds demanded recognition of them from his fellow-men. He
demanded that his works be preserved in *memory*. Based
upon this memory of deeds, a group of men who belonged
together elected one as leader. A kind of regal rank developed. This recognition was even preserved beyond
death. The *memory,* the *remembrance* of the ancestors
or of those who had acquired merit in life, developed.
From this there emerged among some tribes a kind of
religious veneration of the deceased, an *ancestor cult.*
This cult continued into much later times and took the
most varied forms. Among the Rmoahals a man was still
esteemed only to the degree to which he could command
respect at a particular moment through his powers. If
someone among them wanted recognition for what he
had done in earlier days, he had to demonstrate by new
deeds that he still possessed his old power. He had to
recall the old works to memory by means of new ones.
What had been done was not esteemed for its own sake.
Only the second subrace considered the personal character of a man to the point where it took his past life
into account in the evaluation of this character.

A further consequence of memory for the communal
life of man was the fact that groups of men were formed
which were held together by the *remembrance* of common deeds. Previously the formation of groups depended
wholly upon natural forces, upon common descent. Man
did not add anything through his own mind to what
nature had made of him. Now a powerful personality
recruited a number of people for a joint undertaking,
and the memory of this joint action formed a social group.

This kind of social communal life became fully developed only among the third subrace, the Toltecs. It
was therefore the men of this race who first founded what
can be called a community, the first method of forming
a state. The leadership, the government of these communities, was transmitted from one generation to the
next. The father now gave over to the son what previously
survived only in the memory of contemporaries. The
deeds of the ancestors were *not to be forgotten* by their
whole line of descent. What an ancestor had done was
esteemed by his descendants. However, one must realize
that in those times men actually had the power to transmit their gifts to their descendants. Education, after all,
was calculated to mold life through vivid images. The
effectiveness of this education had its foundation in the
personal power which emanated from the educator. He
did not sharpen the power of thought, but in fact, developed those gifts which were of a more instinctive kind.
Through such a system of education the capacities of the
father were generally transmitted to the son.

Under such conditions *personal experience* acquired

more and more importance among the third subrace.
When one group of men separated from another for the
foundation of a new community, it carried along the
remembrance of what it had experienced at the old scene.
But at the same time there was something in this remembrance which the group did not find suitable for itself,
in which it did not feel at ease. Therefore it then tried
something new. Thus conditions improved with every
one of these new foundations. It was only natural that
what was better was imitated. These are the facts which
explain the development of those flourishing communities in the period of the third subrace, described in theo-
sophic literature. The personal experiences which were
acquired found support from those who were *initiated*
into the eternal laws of spiritual development. Powerful
rulers themselves were initiated, so that personal ability
might have full support. Through his personal ability
man slowly prepares himself for initiation. He must first
develop his powers from below in order that the enlightenment from above can be given to him. In this way the
initiated kings and leaders of the Atlanteans came into
being. Enormous power was in their hands, and they
were greatly venerated.

But in this fact also lay the reason for decline and
decay. The development of memory led to the pre-eminent
power of a *personality*. Man wanted to *count for something* through his power. The greater the power became,
the more he wanted to exploit it for himself. The ambition which had developed turned into marked selfishness. Thus the misuse of these powers arose. When one

considers the capabilities of the Atlanteans resulting from their mastery of the life force, one will understand that this misuse inevitably had enormous consequences. A broad power over nature could be put at the service of personal egotism.

This was accomplished in full measure by the fourth subrace, the Primal Turanians. The members of this race, who were instructed in the mastery of the above-mentioned powers, often used them in order to satisfy their selfish wishes and desires. But used in such a manner, these powers destroy each other in their reciprocal effects. It is as if the feet were stubbornly to carry a man forward, while his torso wanted to go backward.

Such a destructive effect could only be halted through the development of a higher faculty in man. This was the faculty of thought. Logical thinking has a restraining effect on selfish personal wishes. The origin of logical thinking must be sought among the fifth subrace, the Primal Semites. Men began to go beyond a mere remembrance of the past and to *compare* different experiences. The faculty of judgment developed. Wishes and appetites were regulated in accordance with this faculty of judgment. One began to *calculate,* to combine. One learned to work with thoughts. If previously one had abandoned oneself to every desire, now one first asked whether thought could approve this desire. While the men of the fourth subrace rushed wildly toward the satisfaction of their appetites, those of the fifth began to listen to an inner voice. This inner voice checks the appetites, although it cannot destroy the claims of the selfish personality.

Thus the fifth subrace transferred the impulses for action to within the human being. Man wishes to come to terms within himself as to what he must or must not do. But what thus was won within, with respect to the faculty of thought, was lost with respect to the control of external natural forces. With this combining thought mentioned above, one can master only the forces of the mineral world, not the life force. The fifth subrace therefore developed thought at the expense of control of the life force. But it was just through this that it produced the germ of the further development of mankind. Now personality, self-love, even complete selfishness could grow freely; for thought alone which works wholly within, and can no longer give direct orders to nature, is not capable of producing such devastating effects as the previously misused powers. From this fifth subrace the most gifted part was selected which survived the decline of the fourth root race and formed the germ of the fifth, the Aryan race, whose mission is the complete development of the thinking faculty.

The men of the sixth subrace, the Akkadians, developed the faculty of thought even further than the fifth. They differed from the so-called Primal Semites in that they employed this faculty in a more comprehensive sense than the former.

It has been said that while the development of the faculty of thought prevented the claims of the selfish personality from having the same devastating effects as among the earlier races, these claims were not destroyed by it. The Primal Semites at first arranged their personal circumstances as their faculty of thought directed. Intelligence took the place of mere appetites and desires. The conditions of life changed. If preceding races were inclined to acknowledge as leader one whose deeds had impressed themselves deeply upon their memory, or who could look back upon a life of rich memories, this role was now conferred upon the *intelligent*. If previously that which lived in a clear remembrance was decisive, one now regarded as best what was most convincing to thought. Under the influence of memory one formerly held fast to a thing until one found it to be inadequate, and in that case it was quite natural that he who was in a position to remedy a want could introduce an innovation. But as a result of the faculty of thought, a fondness for innovations and changes developed. Each wanted to put into effect what his intelligence suggested to him. Turbulent conditions therefore began to prevail under the fifth subrace, and in the sixth they led to a feeling of the need to bring the obdurate thinking of the individual under general *laws*. The splendor of the communities of the third subrace was based on the fact that common memories brought about order and harmony. In the sixth, this order had to be brought about by thought-out laws. Thus it is in this sixth subrace that one must look for the origin of regulations of justice and law.

During the third subrace, the separation of a group of men took place only when they were *forced out* of their community, so to speak, because they no longer felt at ease in the conditions prevailing as a result of memory. In the sixth this was considerably different. The calculating faculty of thought sought the new as such; it spurred men to enterprises and new foundations. The Akkadians

were therefore an enterprising people with an inclina-
tion to colonization. It was commerce, especially, which
nourished the waxing faculty of thought and judgment.

Among the seventh subrace, the Mongols, the faculty
of thought was also developed. But characteristics of
the earlier subraces, especially of the fourth, remained
present in them to a much higher degree than in the fifth
and sixth. They remained faithful to the feeling for memory. And thus they reached the conviction that what is
oldest is also what is most sensible and can best defend
itself against the faculty of thought. It is true that they
also lost the mastery over the life forces, but what developed in them as the thinking faculty also possessed something of the natural might of this life force. Indeed, they
had lost the power over life, but they never lost their
direct, naive *faith* in it. This force had become their *god*,
in whose behalf they did everything they considered right.
Thus they appeared to the neighboring peoples as if
possessed by this secret force, and they surrendered themselves to it in blind trust. Their descendants in Asia and
in some parts of Europe manifested and still manifest
much of this quality.

The faculty of thought planted in men could only attain
its full value in relation to human development when it
received a new impetus in the fifth root race. The fourth
root race, after all, could only put this faculty at the
service of that to which it was educated through the gift
of memory. The fifth alone reached life conditions for
which the proper tool is the ability to think.

iv TRANSITION OF THE FOURTH INTO THE FIFTH ROOT RACE

In this chapter we shall learn about the transition of the fourth, the Atlantean root race, into the fifth, the Aryan, to which contemporary civilized mankind belongs. Only he will understand it aright who can steep himself in the idea of *development* to its full extent and meaning. Everything which man perceives around him is in process of development. In this sense, the use of *thought,* which is characteristic of the men of our fifth root race, had first to develop. It is this root race in particular which slowly and gradually brings the faculty of thought to maturity. In his thought, man decides upon something, and then executes it as the consequence of his own thought. This ability was only in preparation among the Atlanteans. It was not their *own* thoughts, but those which flowed into them from entities of a higher kind, that influenced their will. Thus, in a manner of speaking, their will was directed from outside.

The one who familiarizes himself with the thought of this development of the human being and learns to admit

that man—as earthly man—was a being of a quite different kind in prehistory, will also be able to rise to a conception of the totally different entities which are spoken
of here. The development to be described required enormously long periods of time.

*

What has previously been said about the fourth root
race, the Atlanteans, refers to the great bulk of mankind.
But they followed leaders whose abilities towered far
above theirs. The wisdom these leaders possessed and the
powers at their command were not to be attained by any
earthly education. They had been imparted to them by
higher beings which did not belong directly to earth.
Therefore it was only natural that the great mass of men
felt their leaders to be beings of a higher kind, to be
"messengers" of the gods. For what these leaders knew
and could do would not have been attainable by human
sense organs and by human reason. They were *venerated*
as "divine messengers," and men received their orders,
their commandments, and also their instruction. It was
by beings of this kind that mankind was instructed in
the sciences, in the arts, and in the making of tools. Such
"divine messengers" either directed the communities themselves or instructed men who were sufficiently advanced
in the art of government. It was said of these leaders that
they "communicate with the gods" and were initiated by
the gods themselves into the laws according to which man-

kind had to develop. This was true. In places about
which the average people knew nothing, this initiation,
this communication with the gods, actually took place.
These places of initiation were called temples of the mysteries. From them the human race was directed.

What took place in the temples of the mysteries was
therefore incomprehensible to the people. Equally little
did the latter understand the intentions of their great
leaders. After all, the people could grasp with their senses
only what happened directly upon earth, not what was
revealed from higher worlds for the welfare of earth.
Therefore the teachings of the leaders had to be expressed
in a form unlike communications about earthly events.
The language the gods spoke with their messengers in
the mysteries was not earthly, and neither were the shapes
in which these gods revealed themselves. The higher spirits
appeared to their messengers "in fiery clouds" in order
to tell them how they were to lead men. Only man can
appear in human form; entities whose capacities tower
above the human must reveal themselves in shapes which
are not to be found on earth.

Because they themselves were the most perfect among
their human brothers, the "divine messengers" could receive these revelations. In earlier stages they had already
gone through what the majority of men still had to experience. They belonged among their fellow humans only
in a certain respect. They could assume human form. But
their spiritual-mental qualities were of a superhuman kind.
Thus they were divine-human hybrid beings. One can

also describe them as higher spirits who assumed human bodies in order to help mankind forward on their earthly path. The real home of these beings was not on earth.

These divine-human beings led men, without being able to inform them of the principles by which they directed them. For until the fifth subrace of the Atlanteans, the Primal Semites, men had absolutely no capacities for understanding these principles. The faculty of thought, which developed in this subrace, was such a capacity. But this evolved slowly and gradually. Even the last subraces of the Atlanteans could understand very little of the principles of their divine leaders. They began, at first quite imperfectly, to have a *presentiment* of such principles. Therefore their thoughts and also the laws which we have mentioned among their governmental institutions, were guessed at rather than clearly thought out.

The principal leader of the fifth Atlantean subrace gradually prepared it so that in later times, after the decline of the Atlantean way of life, it could begin a new one which was to be wholly directed by the faculty of thought.

One must realize that at the end of the Atlantean period there existed three groups of man-like beings: 1. The above-mentioned "divine messengers," who in their development were far ahead of the great mass of the people, and who taught divine wisdom and accomplished divine deeds. 2. The great mass of humanity, among which the faculty of thought was in a dull condition, although they possessed natural abilities which modern men have lost.

What are these?

3. A small group of those who were developing the faculty of thought. While they gradually lost the natural abilities of the Atlanteans through this process, they were advancing to the stage where they could grasp the principles of the "divine messengers" with their thoughts.

The second group of human beings was doomed to gradual extinction. The third however could be trained by a being of the first kind to take its direction into its own hands.

From this third group the above-mentioned principal leader, whom occult literature designates as *Manu*, selected the ablest in order to cause a new humanity to emerge from them. These most capable ones existed in the fifth subrace. The faculty of thought of the sixth and seventh subraces had already gone astray in a certain sense and was not fit for further development.

The best qualities of the best had to be developed. This was accomplished by the leader through the isolation of the selected ones in a certain place on earth—in inner Asia—where they were freed from any influence of those who remained behind or of those who had gone astray.

The task which the leader imposed upon himself was to bring his followers to the point where, in their own soul, with their own faculty of thought, they could grasp the principles according to which they had hitherto been directed in a way vaguely sensed, but not clearly recognized by them. Men were to *recognize* the divine forces which they had unconsciously followed. Hitherto the gods had led men through their messengers; now men were to

know about these divine entities. They were to learn to
consider *themselves* as the implementing organs of divine
providence.

The isolated group thus faced an important decision.
The divine leader was in their midst, in human form.
From such divine messengers men had previously received
instructions and orders as to what they were or were not
to do. Human beings had been instructed in the sciences
which dealt with what they could perceive through the
senses. Men had vaguely sensed a divine control of the
world, had felt it in their own actions, but they had not
known anything of it clearly.

Now their leader spoke to them in a completely new
way. He taught them that invisible powers directed what
confronted them visibly, and that they themselves were
servants of these invisible powers, that they had to fulfill
the laws of these invisible powers with their thoughts.
Men heard of the supernatural-divine. They heard that the
invisible spiritual was the creator and preserver of the
visible physical. Hitherto they had looked up to their
visible divine messengers, to the superhuman initiates,
and through the latter was communicated what was and
was not to be done. But now they were considered worthy
of having the divine messenger speak to them of the gods
themselves. Mighty were the words which again and again
he impressed upon his followers: "Until now you have *seen*
those who led you; but there are higher leaders whom you
do not see. It is *these* leaders to whom you are subject. You
shall carry out the orders of the god whom you do not see;
and you shall *obey* one of whom you *can make no image*

to yourselves." Thus did the new and highest commandment come from the mouth of the great leader, prescribing
the veneration of a god whom no sensory-visible image
could resemble, and therefore of whom none was to be
made. Of this great fundamental commandment of the
fifth human root race, the well-known commandment
which follows is an echo: "Thou shalt not make unto thee
any *graven image,* or any likeness of any thing that is in
heaven above, or that is in the earth beneath, or that is in
the water under the earth. . . ." (*Exodus* 20:31).

The principal leader, Manu, was assisted by other divine
messengers who executed his intentions for particular
branches of life and worked on the development of the
new race. For it was a matter of arranging all of life according to the new conception of a divine administration
of the world. Everywhere the thoughts of men were to be
directed from the visible to the invisible. Life is determined by the forces of nature. The course of human life
depends on day and night, on winter and summer, on
sunshine and rain. How these influential visible events
are connected with the invisible, divine powers and how
man was to behave in order to arrange his life in accordance with these invisible powers, was shown to him. All
knowledge and all labor was to be pursued in this sense.
In the course of the stars and of the weather, man was to
see divine decrees, the emanation of divine wisdom. Astronomy and meteorology were taught with this idea. Man
was to arrange his labor, his moral life in such a way that
they would correspond to the wise laws of the divine. Life
was ordered according to *divine commandments,* just as

the *divine thoughts* were explored in the course of the stars and in the changes of the weather. Man was to bring his works into harmony with the dispensations of the gods through *sacrificial acts.*

It was the intention of Manu to direct *everything* in human life toward the higher worlds. *All* human activities, all institutions were to bear a religious character. Through this, Manu wanted to initiate the real task imposed upon the fifth root race. This race was to learn to direct itself by its own thoughts. But such a self-determination can only lead to good if man also places himself at the service of the higher powers. Man should use his faculty of thought, but this faculty of thought should be sanctified by being devoted to the divine.

One can only understand completely what happened at that time if one knows that the development of the faculty of thought, beginning with the fifth subrace of the Atlanteans, also entailed something else. From a certain quarter men had come into possession of knowledge and of arts, which were not *immediately* connected with what the above-mentioned Manu had to consider as his true task. *This* knowledge and these arts were at first devoid of religious character. They came to man in such a way that he could think of nothing other than to place them at the service of self-interest, of his personal needs* . . . To such knowledge belongs for example that of the use of *fire* in human activities. In the first Atlantean times man did not use fire since the life force was available for his

---

* *For the present* it is not permitted to make public communications about the origin of *this* knowledge and *these* arts. A passage from the Akasha Chronicle must therefore be omitted here.

service. But with the passage of time he was less and less in a position to make use of this force, hence he had to learn to make tools, utensils from so-called lifeless objects. He employed fire for this purpose. Similar conditions prevailed with respect to other natural forces. Thus man learned to make use of such natural forces without being conscious of their divine origin. So it was meant to be. Man was not to be *forced* by anything to relate these things which served his faculty of thought, to the divine order of the world. Rather was he to do this *voluntarily* in his thoughts. It was the intention of Manu to bring men to the point where, independently, out of an inner need, they brought such things into a relation with the higher order of the world. Men could *choose* whether they wanted to use the insight they had attained purely in a spirit of personal self-interest or in the religious service of a higher world.

If man was previously forced to consider himself as a link in the divine government of the world, by which for example, the domination over the life force was given to him without his having to use the faculty of thought, he could now employ the natural forces without directing his thoughts to the divine.

Not all the men whom Manu had gathered around him were equal to this decision, but only a few of them. It was from this few that Manu could really form the germ of the new race. He retired with them in order to develop them further, while the others mingled with the rest of mankind.

From this small number of men who finally gathered

around Manu, everything is descended which up to the present, forms the true germs of progress of the fifth root race. For this reason also, two characteristics run through the entire development of this fifth root race. One of these characteristics is peculiar to those men who are animated by higher ideas, who regard themselves as children of a divine universal power; the other belongs to those who put everything at the service of personal interests, of egotism.

The small following remained gathered around Manu until it was sufficiently fortified to act in the new spirit, and until its members could go out to bring this new spirit to the rest of mankind, which remained from the earlier races. It is natural that this new spirit assumed a different character among the various peoples, according to how they themselves had developed in different fields. The old remaining characteristics blended with what the messengers of Manu carried to the various parts of the world. Thus a variety of new cultures and civilizations came into being.

The ablest personalities from the circle around Manu were selected for a gradual direct initiation into his divine wisdom, so that they could become the teachers of the others. A new kind of initiate thus was added to the old divine messengers. It consisted of those who had developed their faculty of thought in an earthly manner just as their fellowmen had done. The earlier divine messengers—and also Manu—had not done this. Their development belonged to higher worlds. They introduced their higher wisdom into earthly conditions. What they gave to mankind was a "gift from above." Before the middle of the Atlantean period men had not reached the point where by their own powers they could *grasp* what the divine decrees were. Now—at the time indicated—they were to attain this point. Earthly thinking was to elevate itself to the concept of the divine. The human initiates united themselves with the divine. This represents an important revolution in the development of the human race. The first Atlanteans did not as yet have a choice as to whether or not they would consider their leaders to be divine messengers. For what the latter accomplished imposed itself as the deed of higher worlds. It bore the stamp of a divine origin. Thus the messengers of the Atlantean period were entities sanctified by their power, surrounded by the splendor which this power conferred upon them. From an external point of view, the human initiates of later times are men among men. But they remain in relation with the higher worlds, and the revelations and manifestations of the divine messengers come to them. Only exceptionally, when a higher necessity arises, do they make use of certain powers which are conferred upon them from above. Then they accomplish deeds which men cannot explain by the laws they know and which therefore they rightly regard as miracles.

But in all this the higher intention is to put mankind on its own feet, fully to develop its faculty of thought. Today the human initiates are the mediators between the people and the higher powers, and only initiation can make one capable of communication with the divine messengers.

The human initiates, the sacred teachers, became leaders of the rest of mankind in the beginning of the fifth root race. The great priest kings of prehistory, who are not spoken of in history, but rather in the world of legend, belong among these initiates. The higher divine messengers retired from the earth more and more, and left the leadership to these human initiates, whom however they assisted in word and deed. Were this not so, man would never attain free use of his faculty of thought. The world is under divine direction, but man is not to be forced to admit this; he is to realize and to understand it by free reflection. When he reaches this point, the initiates will gradually divulge their secrets to him. But this cannot happen all at once. The whole development of the fifth root race is a slow road to this goal. At first Manu himself led his following like children. Then the leadership was gradually transferred to the human initiates. Today progress still consists in a mixture of the conscious and unconscious acting and thinking of men. Only at the end of the fifth root race, when throughout the sixth and seventh subraces a sufficiently great number of men are capable of knowledge, will the greatest among the initiates be able to reveal himself to *them* openly. Then *this* human initiate will be able to assume the principal leadership just as Manu did at the end of the fourth root race. Thus the education of the fifth root race consists in this, that a greater part of humanity will become able freely to follow a human Manu as the germinal race of this fifth root race followed the divine one.

v THE LEMURIAN RACE — *Thru 86 – 71 = 15 pages*

A PASSAGE from the Akasha Chronicle referring to a very distant prehistoric period in the development of mankind, will be set forth in this chapter. This period precedes the one depicted in the descriptions given above. We are here concerned with the *third human root race,* of which it is said in theosophical books that it inhabited the Lemurian Continent. According to these books this continent was situated south of Asia, and extended approximately from Ceylon to Madagascar. What is today southern Asia and parts of Africa also belonged to it.

While all possible care has been taken in the deciphering of the Akasha Chronicle it must be emphasized that nowhere is a dogmatic character to be claimed for these communications. If, to begin with, the reading of things and events so remote from the present is not easy, the translation of what has been seen and deciphered into the language of today presents almost insuperable obstacles.

Dates will be given *later*. They will be better understood when the whole Lemurian period and also the period of our fifth root race up to the present, have been discussed.

The things which are communicated here are surprising
even for the occultist who reads them for the first time—
although the word "surprising" is not quite exact. There·
fore he should only communicate them after the most
careful examination.

The fourth, the Atlantean root race, was preceded by the
so-called *Lemurian*. During its development, events of the
very greatest importance occurred with respect to the
earth and to men. Here, however, something will first be
said of the character of this root race *after* these events, and
only then will the latter be discussed. By and large, *memory* was not yet developed among this race. While men
could have *ideas* of things and events, these ideas did not
remain in the memory. Therefore they did not yet have a
language in the true sense. Rather what they could utter
were natural sounds which expressed their sensations,
pleasure, joy, pain and so forth, but which did not designate external objects.

But their ideas had a quite different strength from those
of later men. Through this strength they acted upon their
environment. Other men, animals, plants, and even lifeless objects could feel this action and could be influenced
purely by ideas. Thus the Lemurian could communicate
with his fellow-men without needing a language. This
communication consisted in a kind of "thought reading."
The Lemurian derived the strength of his ideas directly
from the objects which surrounded him. It flowed to him
from the energy of growth of plants, from the life force
of animals. In this manner he *understood* plants and animals in their inner action and life. He even understood

the physical and chemical forces of lifeless objects in the
same way. When he built something he did not first have
to calculate the load-limit of a tree trunk, the weight of
a stone; he could *see* how much the tree trunk could bear,
where the stone in view of its weight would fit, where it
would not. Thus the Lemurian built without engineering
knowledge on the basis of his faculty of imagination which
acted with the sureness of a kind of instinct. Moreover, to
a great extent, he had power over his own body. When
it was necessary, he could increase the strength of his arm
by a simple effort of the will. For example, he could lift
enormous loads merely by using his will. If later the
Atlantean was helped by his control of the life force, the
Lemurian was helped by his mastery of the will. He was—
the expression should not be misinterpreted—a born magician in all fields of lower human activities.

The goal of the Lemurians was the development of the
will, of the faculty of imagination. The education of
children was wholly directed toward this. The boys were
hardened in the strongest manner. They had to learn to
undergo dangers, to overcome pain, to accomplish daring
deeds. Those who could not bear tortures, who could not
undergo dangers, were not regarded as useful members
of mankind. They were left to perish under these exer-
tions. What the Akasha Chronicle shows with respect to
this raising of children surpasses everything contemporary
man can picture to himself in his boldest imaginings. The
bearing of heat, even of a searing fire, the piercing of the
body with pointed objects, were quite common procedures.

The raising of girls was different. While the female

child was also hardened, everything else was directed toward her developing a strong *imagination*. For example, she was exposed to the storm in order calmly to feel its dreadful beauty; she had to witness the combats of the men fearlessly, filled only with a feeling of appreciation of the strength and power she saw before her. Thereby propensities for dreaming and for fantasy developed in the girl, and these were highly valued. Because no memory existed, these propensities could not degenerate. The dream or fantasy conceptions in question lasted only as long as there was a corresponding external cause. Thus they had a real basis in external things. They did not lose themselves in bottomless depths. It was, so to speak, nature's own fantasy and dreaming which were put into the female soul.

The Lemurians did not have dwellings in our sense, except in their latest times. They lived where nature gave them the opportunity to do so. The caves which they used were only altered and extended insofar as necessary. Later they built such caves themselves and at that time they developed great skill for such constructions. One must not imagine, however, that they did not also execute more artful constructions. But these did not serve as dwellings. In the earliest times they originated in the desire to give to the things of nature a man-made form. Hills were remodeled in such a way that the form afforded man joy and pleasure. Stones were put together for the same purpose, or in order to be used for certain activities. The places where the children were hardened were surrounded with walls of this kind.

Toward the end of this period, the buildings which served for the cultivation of "divine wisdom and divine art" became more and more imposing and ornate. These institutions differed in every respect from what temples were later, for they were educational and scientific institutions at the same time. He who was found fit was here initiated into the science of the universal laws and into the handling of them. If the Lemurian was a born magician, this talent was here developed into art and insight. Only those could be admitted who, through all kinds of discipline, had acquired the ability to overcome themselves to the greatest extent. For all others what went on in these institutions was the deepest secret. Here one learned to know and to control the forces of nature through direct contemplation of them. But the learning was such that in man the forces of nature changed into forces of the will. He himself could thereby execute what nature accomplishes. What later mankind accomplished by reflection, by calculation, at that time had the character of an instinctive activity. But here one must not use the word "instinct" in the same sense in which one is accustomed to apply it to the animal world. For the activities of Lemurian humanity towered high above everything the animal world can produce through instinct. They even stood far above what mankind has since acquired in the way of arts and sciences through memory, reason and imagination. If one were to use an expression for these institutions which would facilitate an understanding of them, one could call them "colleges of will power and of the clairvoyant power of the imagination."

From them emerged the men who, in every respect, became rulers of the others. Today it is difficult to give in words a true conception of all these conditions. For everything on earth has changed since that time. Nature itself and all human life were different, therefore human labor and the relationship of man to man differed greatly from what is customary today.

The air was much thicker even than in later Atlantean times, the water much thinner. And what forms the firm crust of our earth today was not yet as hard as it later became. The world of plants and animals had developed only as far as the amphibians, the birds, and the lower mammals, and as far as vegetable growths which resemble our palms and similar trees. However, all forms were different from what they are today. What now exists only in small forms was then developed to gigantic sizes. At that time our small ferns were trees and formed mighty forests. The modern higher mammals did not exist. On the other hand a great part of humanity was on such a low stage of development that one cannot but designate it as animal. What has been described here was true only of a small part of mankind. The rest lived their life in animalism. In their external appearance and in their way of life these animal men were quite different from the small group. They were not especially different from the lower mammals, which resembled them in form in certain respects.

A few more words must be said about the significance of the above-mentioned temple localities. What was cultivated there was not really religion. It was "divine wisdom and art." Man felt that what was given to him there was a direct gift from the spiritual universal forces. When he received this gift he considered himself a "servant" of these universal forces. He felt himself "sanctified" from everything unspiritual. If one wishes to speak of religion at this stage of the development of mankind, one could call it "religion of the will." The religious temper and dedication lay in the fact that man guarded the powers granted to him as a strict, divine "secret," and that he led a life through which he sanctified his power. Persons who had such powers were regarded by others with great awe and veneration. And this awe and veneration were not called forth by laws or something similar, but by the immediate power which these persons exercised. The uninitiated of course stood under the magical influence of the initiated. It was also natural that the latter considered themselves to be sanctified personages. For in their temples they participated in direct contemplation of the active forces of nature. They looked into the creative workshop of nature. They experienced a communion with the beings which build the world itself. One can call this communication an association with the gods. What later developed as "initiation," as "mystery," emerged from this original manner of communication of men with the gods. In subsequent times this communication had to become different, since the human imagination, the human spirit, took other forms.

Of special importance is something which occurred in the course of Lemurian development by virtue of the fact that the women lived in the manner described above. They thereby developed special human powers. Their faculty of imagination which was in alliance with nature, became the basis for a higher development of the life of ideas. They took the forces of nature into themselves, where they had an after-effect in the soul. Thus the germs of memory ✕ were formed. With memory was also born the capacity to form the first and simplest moral concepts.

The development of the will among the male element at first knew nothing of this. The man followed instinctively either the impulses of nature or the influences emanating from the initiated.

It was from the manner of life of the women that the first ideas of "good and evil" arose. There one began to love some of the things which had made a special impression on the imagination, and to abhor others. While the control which the male element exercised was directed more toward the external action of the powers of the will, toward the manipulation of the forces of nature, beside it in the female element there developed an action through the soul, through the inner, personal forces of man. The development of mankind can only be correctly understood by the one who takes into consideration that the first progress in the life of the imagination was made by women. The development connected with the life of the imagination, with the formation of memory, of customs which formed the seeds for a life of law, for a kind of morals, came from this side. If man had seen and exercised the forces of nature, woman became the first *interpreter* of them. It was a special new manner of living through reflection which developed here. This manner had something much more personal than that of the men. One must imagine this manner of the women to have been also a kind of clairvoyance, although it differed from the magic of the will of the men. In her soul woman was accessible to another kind of spiritual powers. The latter spoke more to the feeling element of the soul, less to the spiritual, to which man was subject. Thus there emanated from men an effect which was more natural-divine, from women one which was more soul-divine.

The development which woman went through during the Lemurian period had the result that at the appearance of the next—the Atlantean—root race on earth, an important role devolved upon her. This appearance took place under the influence of highly developed entities, who were familiar with the laws of the formation of races and capable of guiding the existing forces of human nature into such paths that a new race could come into being. These beings will be specially mentioned further on. May it suffice for the moment to say that they possessed superhuman wisdom and power. They now isolated a small group out of Lemurian mankind and designated these to be the ancestors of the coming Atlantean race. The place where they did this was situated in the tropical zone. Under their direction the men of this group had been trained in the control of the natural forces. They were very strong, and knew how to win the most diverse treasures from the earth. They could cultivate the fields and use their fruits for their subsistence. They had become characters of strong will through the discipline to which they had been subjected. Their souls and hearts were developed only in small measure. On the other hand these had been developed among the women. Memory and fantasy and everything connected with them were to be found among the latter.

The above-mentioned leaders caused the group to divide itself into smaller groups. They put the women in charge of ordering and establishing these groups. Through her memory, woman had acquired the capacity to make the experiences and adventures of the past useful for the future. What had proved helpful yesterday she used today and realized that it would also be useful tomorrow. The institutions for communal life therefore emanated from her. Under her influence the concepts of "good and evil" developed. Through her thoughtful life she had acquired an understanding for nature. Out of the observation of nature, those ideas developed in her according to which she directed the actions of men. The leaders had arranged things in such a way that through the *soul* of woman, the willful nature, the vigorous strength of man were ennobled and refined. Of course one must represent all this to oneself as childish beginnings. The words of our language all too easily call up ideas which are taken from the life of the present.

By way of the awakened soul life of the women the leaders first developed the soul life of the men. In the colony we have described, the influence of the women was therefore very great. One had to go to them for advice when one wanted to interpret the signs of nature. The whole manner of their soul life however was still dominated by the "hidden" human soul forces. One does not describe the matter quite exactly, but fairly closely, if one speaks of a somnambulistic contemplating among these women. In certain higher dreams the secrets of nature were divulged to them and they received the impulses for their actions. Everything was animated for them and showed itself to them in soul powers and apparitions. They abandoned themselves to the mysterious weaving of their soul forces. That which impelled them to their actions were "inner voices," or what plants, animals, stones, wind and clouds, the whispering of the trees, and so on, told them.

From this state of soul originated that which one can call human religion. The spiritual in nature and in human life gradually came to be venerated and worshipped. Some women attained a special preeminence because out of special mysterious depths they could interpret what the world contained.

Thus it could come to pass among such women that that which lived within them could transpose itself into a kind of natural language. For the beginning of language lies in something which is similar to song. The energy of thought was transformed into audible sound. The inner rhythm of nature sounded from the lips of "wise" women. One gathered around such women and in their songlike sentences felt the utterances of higher powers. Human worship of the gods began with such things.

For that period there can be no question of "sense" in that which was spoken. Sound, tone, and rhythm were perceived. One did not imagine anything along with these, but absorbed in the soul the power of what was heard.

The whole process was under the direction of the higher leaders. They had inspired the "wise" priestesses with tones and rhythms in a manner which cannot now be further discussed. Thus they could have an ennobling effect on the souls of men. One can say that in this way the true life of the soul first awakened.

In this realm, beautiful scenes are shown by the Akasha Chronicle. One of these will be described. We are in a forest, near a mighty tree. The sun has just risen in the east. The palmlike tree, from around which the other trees have been removed, casts mighty shadows. The priestess, her face turned to the east, ecstatic, sits on a seat made of rare natural objects and plants. Slowly in rhythmical sequence, a few strange, constantly repeated sounds stream from her lips. A number of men and women are sitting in circles around her, their faces lost in dreams, absorbing inner life from what they hear.

Other scenes too can be seen. At a similarly arranged place a priestess "sings" in a similar manner, but her tones have in them something mightier, more powerful. Those around her move in rhythmic dances. For this was the other way in which "soul" entered into mankind. The mysterious rhythms which one had heard from Nature were imitated by the movements of the limbs. One thereby felt *at one* with nature and with the powers acting in her.

The place on earth in which this stock of a coming race of men was developed was especially suited for this purpose. It was one where the then still turbulent earth had become fairly calm. For Lemuria was turbulent. After all, the earth at that time did not yet have its later density. The thin ground was everywhere undermined by volcanic forces which broke forth in smaller or larger streams. Mighty volcanoes existed almost everywhere and developed a continuous destructive activity. Men were accustomed to reckoning with this fiery activity in everything they did. They also used this fire in their labors and contrivances. Their occupations were often such that the fire of nature served as a basis for them in the same way as artificial fire does in human labor today.

It was through the activity of this volcanic fire that the destruction of the Lemurian land came about. While the part of Lemuria from which the parent race of the Atlanteans was to develop had a hot climate, it was by and large free of volcanic activity.

Human nature could unfold more calmly and peacefully here than in the other regions of the earth. The more nomadic life of former times was abandoned, and fixed settlements became more and more numerous.

One must represent to oneself that at that time the human body still had very malleable and pliant qualities. This body still changed form whenever the inner life changed. Not long before, men had still been quite diverse as regards their external form. At that time the external influence of region and climate were still decisive in respect to their form. Only in the colony described did the body of man increasingly become an expression of his inner soul life. Moreover, this colony had an advanced, externally more nobly formed race of men. One must say that through the things which they

had done, the leaders had really first created what is the
true human form. This occurred quite slowly and gradually. It happened in such a way that the soul life of
man was first developed and that the still soft and malleable body adapted itself to this. It is a law in the development of mankind that, as progress continues, man has
less and less of a molding influence on his physical body.
This physical human body in fact received a fairly un-
changing form only with the development of the faculty
of reason and with the hardening of the rock, mineral,
and metal formations of earth connected with this development. For in the Lemurian and even in the Atlantean period, stones and metals were much softer than
later.

This is not contradicted by the fact that there exist
descendants of the last Lemurians and Atlanteans who
today exhibit forms as fixed as the human races which
were formed later. These remnants had to adapt themselves to the changed environmental conditions of earth
and thus became more rigid. Just this is the reason for
their decline. They did not transform themselves from
within; instead, their less developed interior was forced
into rigidity from the outside and thus compelled to
stagnation. This stagnation is really a regression, for the
inner life, too, has degenerated because it could not fulfill
itself within the rigid external bodily structure.

Animal life was subject to even greater changeability.
We shall speak further about the animal species existing
at the time of the development of man and about their

origin, as well as about the development of new animal
forms after man already existed. Here we shall say only
that the existing animal species continually transformed
themselves and that new ones were developing. This
transformation was of course a gradual one. The reasons
for the transformation lay in part in a change of habitat
and of the manner of life. The animals had a capacity of
extraordinarily rapid adaptation to new conditions. The
malleable body changed its organs comparatively rapidly,
so that after a more or less brief period the descendants
of a particular animal species resembled their ancestors
only slightly. The same was the case in even greater measure for the plants. The greatest influence on the transformation of men and animals was exercised by man
himself. This was true whether he instinctively brought
organisms into such an environment that they assumed
certain forms, or whether he achieved this by experiments
in breeding. The transforming influence of man on nature was immeasurably great at that time, compared
with the conditions of today. This was especially the
case in the colony we have described. For there the
leaders directed this transformation in a way of which
men were not conscious. This was the case to such a
degree that when men left the colony in order to found
the different Atlantean races, they could take with them
a highly developed knowledge of the breeding of animals and plants. The labor of cultivation in Atlantis was
then essentially a consequence of the knowledge thus
brought along. But here again it must be emphasized

that this knowledge had an instinctive character. In this
state essentially it remained among the first Atlantean
races.

The preeminence of the feminine soul, which has
been described, was especially strong in the last Lemurian period and continued into the Atlantean times, during which the fourth subrace was preparing itself. But
one must not imagine that this was the case among all
of mankind. It was true, however, for that part of the
population of earth from which the truly advanced races
later emerged. This influence exercised the strongest effect upon all that which in man is "unconscious." The
development of certain constant gestures, the refinements
of sensory perception, the feeling for beauty, a good part
of the general life of sensations and feelings which is
common to all men—all this originally emanated from
the spiritual influence of woman. It is not an over-state-
ment if one interprets the reports of the Akasha Chronicle in such a way as to affirm, "The civilized nations
have a bodily form and expression, as well as certain
bases of physical-soul life, which were imprinted upon
them by woman."

In the next chapter we shall go back to earlier periods
of the development of mankind, during which the population of earth still belonged to only one sex. The development of the two sexes will then be described.

The spiritual influence of woman

To here ...

vi THE DIVISION INTO SEXES

MUCH AS THE HUMAN FORM in those ancient times described in the preceding chapters differed from the form of present-day man, one comes to conditions still more dissimilar if one goes even further back in the history of mankind. For only in the course of time did the forms of man and woman develop from an older, basic form in which human beings were neither the one nor the other, but rather were both at once. He who wants to form an idea of these enormously distant periods of the past must however liberate himself completely from the habitual conceptions taken from what man sees around him.

The times into which we now look back lie somewhat before the middle of the epoch which in the preceding passages was designated as the Lemurian. At that time the human body still consisted of soft and malleable materials. The other forms of earth also were still soft and malleable. As opposed to its later hardened condition, earth was still in a welling, more fluid one. As the human soul at that time embodied itself in matter, it could adapt this matter to itself in a much greater degree than later. That the soul takes on a male or a female body is due to the fact that the development of external terrestrial nature forces the one or the other upon it. While the material substances had not yet become rigid, the soul could force these substances to obey its own laws. It made of the body an impression of its own nature. But when matter became denser the soul had to submit to the laws impressed upon this matter by external terrestrial nature. As long as the soul could still control matter, it formed its body as neither male nor female, but, instead, gave it qualities which embraced both at the same time. For the soul is simultaneously male and female. It carries these two natures in itself. Its male element is related to what is called *will,* its female element to what is called *imagination.*

The external formation of earth resulted in that the body assumed a one-sided form. The male body has taken a form which is conditioned by the element of will; the female body on the other hand, bears the stamp of imagination. Thus it comes about that the two-sexed, male-female soul inhabits a single-sexed, male *or* female body. In the course of development the body had taken a form determined by the external terrestrial forces, so that it was no longer possible for the soul to pour its whole inner energy into this body. The soul had to retain something of this energy within itself and could let only a part of it flow into the body.

If one continues with the Akasha Chronicle, the following becomes apparent. In an ancient period, human forms appear before us which are soft, malleable and quite different from later ones. They still carry the nature of man and woman within themselves to an equal degree. In the course of time, the material substances become denser; the human body appears in two forms, one of which begins to resemble the subsequent shape of man, the other that of woman. When this difference had not yet appeared, every human being could produce another human being out of himself. Impregnation was not an external process, but was something which took place inside the human body itself. By becoming male or female, the body lost this possibility of self-impregnation. It had to act together with another body in order to produce a new human being.

Note:

The division into sexes takes place when the earth enters a certain stage of its densification. The density of matter inhibits a portion of the force of reproduction. That portion of this force which is still active needs an external complementation through the opposite force of another human being. The soul however must retain a portion of its earlier energy within itself, in man as well as in woman. It cannot use this portion in the physical external world.

This portion of energy is now directed toward the interior of man. It cannot emerge toward the exterior; therefore it is freed for inner organs.

Here an important point in the development of mankind appears. Previously that which is called spirit, the faculty of thought, could not find a place in man. For this faculty would have found no organs for exercising its functions. The soul had employed all its energy to-

ward the exterior, in order to build up the body. But
now the energy of the soul, which finds no external em-
ployment, can become associated with the spiritual energy,
and through this association those organs are developed
in the body which later make of man a thinking being.
Thus man could use a portion of the energy which previously he employed for the production of beings like
himself, in order to perfect his own nature. The force by
which mankind forms a thinking brain for itself is the
same by which man impregnated himself in ancient times.
The price of thought is single-sexedness. By no longer
impregnating themselves, but rather by impregnating each
other, human beings can turn a part of their productive
energy within, and so become thinking creatures. Thus
the male and the female body each represent an imper-
fect external embodiment of the soul, but thereby they
become more perfect creatures inwardly.

This transformation of man takes place very slowly
and gradually. Little by little, the younger, single-sexed
male or female forms appear beside the old double-sexed
ones.

It is again a kind of fertilization which takes place in
man when he becomes a creature endowed with spirit.
The inner organs which can be built up by the surplus
soul energy are fructified by the spirit. In itself the soul
is two-sided: male-female. In ancient times it also formed
its body on this basis. Later it can form its body only in
such a way that for the external it acts together with
another body; thereby the soul itself receives the capacity to act together with the spirit. For the external, man

is henceforward fertilized from the outside, for the in-
ternal, from the inside, through the spirit. One can say
that the male body now has a female soul, the female
body a male soul. This inner one-sidedness of man is
compensated by fertilization through the spirit. The one-
sidedness is abolished. Both the male soul in the female
body and the female soul in the male body again become
double-sexed through fructification by the spirit. Thus
man and woman are different in their external form;
internally their spiritual one-sidedness is rounded out to
a harmonious whole. Internally, spirit and soul are fused
into one unit. Upon the male soul in woman the action
of the spirit is female, and thus renders it male-female;
upon the female soul in man the action of the spirit is
male, and thus renders it male-female also. The double-
sexedness of man has retired from the external world,
where it existed in the pre-Lemurian period, into his
interior.

One can see that the higher inner essence of a human
being has nothing to do with man or woman. The inner
equality, however, does result from a male soul in woman,
and correspondingly from a female soul in man. The
union with the spirit finally brings about the equality;
but the fact that before the establishment of this equality there exists a difference involves a *secret* of human
nature. The understanding of this secret is of great significance for all mystery science. It is the key to important enigmas of life. *For the present we are not
permitted to lift the veil which is spread over this secret. . . .*
Thus physical man has developed from double-sexed-

ness to single-sexedness, to the separation into male and female. In this way man has become a spiritual being of the kind which he is now. But one must not suppose that no beings which possessed cognition had been in contact with the earth before then. When one follows the Akasha Chronicle it does indeed appear that in the first Lemurian period, later physical man, because of his double sex, was a totally different being from that which one today designates as man. He could not connect any sensory perceptions with thoughts; he did not think. His life was one of impulses. His soul expressed itself only in instincts, in appetites, in animal desires and so on. His consciousness was *dreamlike;* he lived in *dullness*.

But there were other beings among these men. These of course were also double-sexed. For at the stage of terrestrial development of that time no male or female human body could be produced. The external conditions did not yet exist for this. But there were other beings which could acquire knowledge and wisdom in spite of their double-sexedness. This was possible because they had gone through a quite different development in a still more remote past. It was possible for their soul to be fructified by the spirit without first awaiting the development of the inner organs of the physical body of man. By means of the physical brain, the soul of contemporary man can think only that which it receives from the outside through the physical senses. This is the condition to which the development of man's soul has led. The human soul had to wait until a brain existed which became the

mediator with the spirit. Without this detour, *this* soul
would have remained spiritless. It would have remained
arrested at the stage of dreamlike consciousness. This
was different among the superhuman beings mentioned
above. In previous stages their soul had developed organs
which needed nothing physical in order to enter into
contact with the spirit. Their knowledge and wisdom were
supersensibly acquired. Such knowledge is called intuitive. Contemporary man attains such intuition only at a
later stage of his development; this intuition makes it
possible for him to enter into contact with the spirit
without sensory mediation. He must make a detour
through the world of sensory substance. This detour is
called the descent of the human soul into matter, or
popularly, "the fall of man."

Because of a different earlier development, the superhuman beings did not have to take part in this descent.
Since their soul had already attained a higher stage, their
consciousness was not dreamlike, but inwardly clear. Their
acquisition of knowledge and wisdom was a *clairvoyance*
which had no need of senses or of an organ of thought.
The wisdom according to which the world is built shone
into their soul directly. Therefore they could become the
leaders of youthful humanity which was still sunk in
dullness. They were the bearers of a "primeval wisdom,"
toward the understanding of which mankind is only now
struggling along the detour mentioned above. They differed from what one calls "man" through the fact that
wisdom shone upon them as the sunlight does upon us,

as a free gift "from above." "Man" was in a different
position. He had to acquire wisdom by the work of the
senses and of the organ of thought. Originally it did not
come to him as a free gift. He had to *desire* it. Only when
the *desire* for wisdom lived in man, did he acquire it
through his senses and his organ of thought. Thus a new
impulse had to awaken in the soul: the desire, the *longing for knowledge*. In its earlier stages the human soul
could not have had this longing. The impulses of the
soul were directed only toward materialization in that
which assumed form externally—in what took place in it
as a dreamlike life—but not toward cognition of the external world, nor toward knowledge. It is with the division into sexes that the impulse toward knowledge first
appears.

The superhuman beings received wisdom by way of
clairvoyance just because they did not have this desire
for it. They waited until wisdom shone into them, as we
wait for the sunlight, which we cannot produce at night,
but which must come to us by itself in the morning.

The longing for knowledge is produced by the fact that
the soul develops inner organs, the brain and so forth,
by means of which it gains possession of knowledge. This
is a consequence of the circumstance that a part of the
energy of the soul is no longer directed toward the outside, but toward the inside. The superhuman beings
however, which have not carried out this separation of
their spiritual forces, direct all the energy of their soul
toward the outside. Therefore that force is also available
to them externally for fructification by the spirit, which

"man" turns inward for the development of the organs
of cognition.

Now that force by means of which one human being
turns toward the outside in order to act together with
another is *love*. The superhuman beings directed all their
love outward in order to let universal wisdom flow into
their soul. "Man" however can only direct a part of it
outward. "Man" became sensual, and thereby his love
became sensual. He draws away from the outside world
that part of his nature which he directs toward his inner
development. And thus that arises which one calls *selfishness*. When he became man or woman in the physical
body, "man" could surrender himself with only a part
of his being; with the other part he separated himself
from the world around him. He became selfish. And his
action toward the outside became selfish; his striving after
inner development also became selfish. He loved because
he *desired*, and likewise he thought because he *desired*
wisdom.

The selfless, all-loving natures, the leaders, the superhuman beings, confronted man, who was still childishly
selfish.

The soul, which among these beings does not reside
in a male or female body, is itself male-female. It loves
without *desire*. Thus the innocent soul of man loved before the division into sexes, but at that time it could not
understand, because it was still at an inferior stage, that
of dream consciousness. The soul of the superhuman
beings also loves in this manner, however, with *understanding* because of its advanced development. "Man"

must pass through selfishness in order to attain selflessness again at a higher stage, where, however, it will be combined with completely clear consciousness.

The task of the superhuman natures, of the great leaders, was that they impressed upon youthful man their own character, that of *love*. They could do this only for that part of the spiritual energy which was directed outward. Thus *sensual love* was produced. It is therefore a consequence of the activity of the soul in a male or female body. Sensual love became the force of physical human development. This love brings man and woman together insofar as they are physical beings. Upon this love rests the progress of physical humanity.

It was only over this love that the superhuman natures had power. That part of human soul energy which is directed inward and is to bring about cognition by the detour through the senses—that part is withdrawn from the power of *those* superhuman beings. However, they themselves had never descended to the development of corresponding inner organs. They could clothe the impulse toward the external in love, because love acting toward the external was part of their own nature. Because of this, a gulf opened between them and youthful mankind. Love, at first in sensual form, they could plant in man; knowledge they could not give, for their own knowledge had never made the detour through the inner organs which man was now developing. They could speak no language which a creature with a brain could have understood.

The inner organs of man mentioned above first became

ripe for a contact with the spirit only at that stage of
terrestrial existence which lies in the middle of the Lemurian period; but they had already been formed incom-
pletely, at a much earlier stage of development. For the
soul had already gone through physical embodiments in
preceding times. It had lived in dense substance, not on
earth but on other celestial bodies. Details about this
must be given later. At present we shall say only that the
terrestrial beings previously lived on another planet, where,
in accordance with the prevailing conditions, they developed up to the point at which they were when they
arrived on earth. They put off the substances of this
preceding planet like clothing and, at the level of development which they thus attained, became pure soul germs
with the capacity to perceive, to feel and so forth—in short,
to lead that dreamlike life which remained peculiar to
them in the first stages of their terrestrial existence.

The superhuman entities previously mentioned, the
leaders in the field of love, had already been so perfect
on the preceding planet that they did not have to descend
to develop the rudiments of those inner organs.

But there were other beings, not as far advanced as
these leaders of love, who on the preceding planet were
still numbered among "men," but at that period were
hurrying ahead of men. Thus, at the beginning of the
formation of the earth, they were further advanced than
men, but still were at the stage where knowledge must be
acquired through inner organs. These beings were in a
special position. They were too far advanced to pass
through the physical human body, male or female, but

on the other hand, were not so far advanced that they could act through full clairvoyance like the leaders of love. They could not yet be *beings of love;* they could no longer be "men." Thus they could only continue their own development as half superhuman beings, in which they were aided by men. They could speak to creatures with a brain in a language which the latter could understand. Thereby the human soul energy which was turned inward was stimulated, and could connect itself with knowledge and wisdom. It was thus that wisdom of a human kind first appeared on earth. The "half superhuman beings" mentioned above could use this human wisdom in order to achieve for themselves that of perfection which they still lacked. In this manner they became the stimulators of human wisdom. One therefore calls them *bringers of light* (Lucifer). Youthful mankind thus had two kinds of leaders: beings of love and beings of wisdom. Human nature was balanced between love and wisdom when it assumed its present form on this earth. By the beings of love it was stimulated to physical development, by the beings of wisdom to the perfection of the *inner* nature. As a consequence of physical development, humanity advances from generation to generation, forms new tribes and races; through inner development individuals grow toward inner perfection, become knowing and wise men, artists, technicians etc. Physical mankind strides from race to race; each race hands down its sensorily perceptible qualities to the following one through physical development. Here the law of heredity holds sway. The children carry within themselves the physical

characteristics of the fathers. Beyond this lies a process of
spiritual-soul perfection which can only take place through
the development of the soul itself.

With this we stand before the law of the development
of the soul within terrestrial existence. This development
is connected with the law and mystery of *birth* and *death*.

vii THE LAST PERIODS BEFORE
THE DIVISION INTO SEXES

We SHALL NOW DESCRIBE the state of man before his division into male and female. At that time the body consisted of a soft malleable mass. The will had a much greater power over this mass than later. When man separated from his parent entity he appeared as a truly articulated organism, but as an incomplete one. The further development of the organs took place outside the parent entity. Much of what later matured inside the mother organism was at that time brought to completion outside of it by a force which was akin to our will power. In order to bring about such an external maturation the care of the parent being was necessary. Man brought certain organs into the world which he later cast off. Others, which were quite incomplete at his first appearance, developed more fully. The whole process had something which can be compared with the emergence from an egg-form and the casting off of an eggshell, but here one must not think of a firm eggshell.

The body of man was warm-blooded. This must be

stated explicitly, for in even earlier times it was different, as will be shown later. The maturation which took place outside the mother organism occurred under the influence of an increased warmth which also was supplied from the outside. But one must by no means think that the egg-man—as he will be called for the sake of brevity—was *brooded*. The conditions of heat and fire on the earth of that time were different from those of later times. By means of his powers man could confine fire, or respectively, heat, to a certain space. He could, so to speak, contract, (concentrate) heat. He was thus in a position to supply the young organism with the warmth which it needed for its maturation.

The most highly developed organs of man at that time were the organs of motion. The sense organs of today were as yet quite undeveloped. The most advanced among them were the organs of hearing and of perception of cold and hot, the sense of touch; the perception of light lagged far behind. Man came into the world with the senses of hearing and touch; the perception of light developed somewhat later.

Everything which is said here applies to the last periods before the division into sexes. This division took place slowly and gradually. Long before its actual occurrence, human beings were already developing in such a way that one individual would be born with more male, another with more female characteristics. Each human being however also possessed the opposite sexual characteristics, so that self-impregnation was possible. But the latter could not always take place, because it depended on the influences of external conditions in certain seasons. With respect to many things and to a great extent, man was generally dependent on such outer conditions. Therefore he had to regulate all his institutions in accordance with such external conditions, for example, in accordance with the course of the sun and the moon. But his regulation did not take place consciously in the modern sense, but was accomplished in a manner which one must call instinctive. With this we already indicate the soul life of man of that time.

This soul life cannot be described as a true inner life. Physical and soul activities and qualities were not yet strictly separated. The outer life of nature was still experienced by the soul. Each single disturbance in the environment acted powerfully on the sense of hearing especially. Every disturbance of the air, every movement was "heard." In their movements wind and water spoke an "eloquent language" to man. In this manner a perception of the mysterious activity of nature penetrated into him. This activity reverberated in his soul. His own activity was an echo of these impressions. He transformed the perceptions of sound into his own activity. He lived among such tonal movements and expressed them by his will. In this way he was impelled to all his daily labors.

He was influenced in a somewhat lesser degree by the influences which act upon the touch. But they also played an important role. He "felt" the environment in his body and acted accordingly. From such influences upon the touch he could tell when and how he had to work. He knew from them where he should rest. In them he recognized and avoided dangers which threatened his life. In accordance with these influences he regulated his food intake.

The remainder of the soul life took its course in a manner quite different from that of later periods. In the soul lived *images* of external objects, not conceptions of them. For instance, when man entered a warmer space from a colder one, a certain colored image arose in his soul. But this colored image had nothing to do with any external object. It originated in an inner force which was akin to the will. Such images continuously filled the soul. One can compare this only with the flowing dream impressions of man. At that time the images were not completely irregular, but proceeded according to law. Therefore, in relation to this stage of mankind, one should speak of an image consciousness rather than of a dream consciousness. For the most part, colored images filled this consciousness. But these were not the only kind. Thus man wandered through the world, and through his hearing and touch participated in the events of this world; but in his soul life this world was mirrored in images which were very unlike what existed in the external world. Joy and sorrow were associated with the images of the soul to a much lesser degree than is the case today with the ideas of men which reflect their perceptions of the external world. It is true that one image awakened happiness, another displeasure, one hate, another love; but these feelings had a much paler character.

On the other hand, strong feelings were aroused by something else. At that time man was much more active

than later. Everything in his environment as well as the
images in his soul, stimulated him to activity, to movement. When his activity could proceed without hindrance,
he experienced pleasure, but when this activity was hindered in any way, he felt displeasure and discomfort. It
was the absence or presence of hindrances to his will which
determined the content of his sensations, his joy and his
pain. This joy, or this pain were again released in his
soul in a world of living images. Light, clear, beautiful
images lived in him when he could be completely free in
his actions; dark, misshapen images arose in his soul when
his movements were hindered.

Until now the average man has been described. Among
those who had developed into a kind of superhuman
beings, (cf. page 96) soul life was different. Their soul
life did not have this instinctive character. Through their
senses of hearing and touch they perceived deeper mysteries of nature, which they could interpret consciously.
In the rushing of the wind, in the rustling of the trees,
the laws, the *wisdom* of nature were unveiled to them.
The images in their souls did not merely represent reflections of the external world, but were likenesses of the
spiritual powers of the world. They did not perceive
sensory objects, but spiritual entities. For example, the
average man experienced fear, and an ugly, dark image
arose in his soul. By means of such images the superhuman being received information and revelation about
the spiritual entities of the world. The processes of nature
did not appear to him as dependent on lifeless natural
laws, as they do to the scientist of today, but rather as

the actions of spiritual beings. External reality did not yet exist, for there were no external senses. But spiritual reality was accessible to the higher beings. The spirit shone into them as the sun shines into the physical eye of a man today. In these beings, cognition was what one may call intuitive knowledge in the fullest sense of the word. For them there was no combining and speculating, but an immediate perception of the activity of spiritual beings. Therefore, these superhuman individuals could receive communications from the spiritual world directly into their will. They consciously directed the other men. They received their mission from the world of spirits and acted accordingly.

When the time came in which the sexes separated, these beings considered it their task to act upon the new life in accordance with their mission. The regulation of sexual life emanated from them. Everything which relates to the reproduction of mankind originated with them. In this they acted quite consciously, but the other men could only feel this influence as an instinct implanted in them. Sexual love was implanted in man by immediate transference of thought. At first all its manifestations were of the noblest character. Everything in this area which has taken on an ugly character comes from later times, when men became more independent and when they corrupted an originally pure impulse. In these older times there was no satisfaction of the sexual impulse for its own sake. Then, everything was a sacrificial service for the continuation of human existence. Reproduction was regarded as a sacred matter, as a service which man owes

to the world. Sacrificial priests were the directors and
regulators in this field.

Of a different kind were the influences of the half
superhuman beings (cf. page 96/97). The latter were
not developed to the point of being able to receive the
revelations of the spiritual world in an entirely pure form.
Along with these impressions of the spiritual world, the
effects of the sensible earth also arose among the images
of their souls. The truly superhuman beings received
no impressions of joy and pain through the external world.
They were wholly given over to the revelations of the
spiritual powers. Wisdom flowed to them as light does to
sensory beings; their will was directed toward nothing
but acting in accordance with this wisdom. In this acting
lay their highest joy. Wisdom, will, and activity constituted their nature. This was different among the half
superhuman entities. They felt the impulse to receive
impressions from the outside, and with the satisfaction
of this impulse they connected joy, with its frustration,
displeasure. Through this they differed from the superhuman entities. To these entities, external impressions
were nothing but confirmations of spiritual revelations.
They could look out into the world without receiving
anything more than a reflection of what they had already
received from the spirit. The half-superhuman beings
learned something new, and therefore *they* could become
leaders of men when in human souls mere images changed
into likenesses and conceptions of external objects. This
happened when a portion of the previous reproductive
energy of man turned inward, at the time when entities

with brains were developed. With the brain man also received the capacity to transform external sensory impressions into conceptions.

It must therefore be said that by half-superhuman beings man was brought to the point of directing his inner nature toward the sensuous external world. He was not permitted to open the images of his soul directly to pure spiritual influences. The capacity of perpetuating the existence of his kind was implanted in him as an instinctive impulse by superhuman beings. Spiritually, he would at first have had to continue a sort of dream existence if the half-superhuman beings had not intervened. Through their influence the images of his soul were directed toward the sensuous, external world. He became a being which was conscious of itself in the world of the senses. Thereby it came about that man could consciously direct his actions in accordance with his perceptions of the world of the senses. Before this he had acted from a kind of instinct. He had been under the spell of his external environment and of the powers of higher individualities, which acted on him. Now he began to follow the impulses and enticements of his conceptions. Therewith free choice became possible for man. This was the beginning of "good and evil."

Before we continue in this direction, something will be said concerning the environment of man on earth. In addition to man there existed animals, which, for *their* kind, were at the same stage of development as he. According to current ideas one would include them among the reptiles. Apart from them, lower forms of animal life existed. Between man and the animals there was an essential difference. Because of his still malleable body, man could live only in those regions of the earth which had not yet passed over into the most solid material form. And in these regions animal organisms which had a similarly plastic body lived with him. But in other regions lived animals which already had dense bodies and also had developed separate sexedness and the senses. Where they had come from, will be explained later. These animals could not develop further because their bodies had taken on this denser materiality too soon. Some species of these became extinct, others have perpetuated their kind to the point of contemporary forms. Man could attain higher forms because he remained in the regions which corresponded to his state at that time. Thereby his body remained so pliant and soft that he could develop the organs which were to be fructified by the spirit. With this development his external body had reached the point where it could pass over into denser materiality and become a protective envelope for the more delicate spiritual organs.

Not all human bodies, however, had reached this point. There were few advanced ones. These were first animated by spirit. Others were not animated. If the spirit had penetrated into them it could have developed only in a defective manner because of the as yet incomplete inner organs. Therefore, at first these human beings were compelled to develop further without spirit. A third kind had reached the point where weak spiritual impulses could act in them. They stood between the two other

kinds. Their mental activity remained dull. They had
to be led by higher spiritual powers. All possible transi-
tions existed between these three kinds. Further development was now possible only in that a portion of the human
beings attained higher forms at the expense of the others.
First, the completely mindless ones had to be abandoned.
A mingling with them for the purpose of reproduction
would have pulled the more highly developed down to
their level. Everything which had been given a mind was
therefore separated from them. Thereby the latter descended more and more to the level of animalism. Thus,
alongside man there developed manlike animals. Man
left a portion of his brothers behind on his road in order
that he himself might ascend higher. This process had
by no means come to an end. Among the men with a dull
mental life those who stood somewhat higher could advance only if they were raised to an association with
higher ones, and separated themselves from those less
endowed with spirit. Only thus could they develop bodies
which would be fit to receive the full human spirit. After
a certain time the physical development had come to a
kind of stopping-point, in that everything which lay above
a certain boundary remained human. Meanwhile, the
conditions of life on earth had changed in such a way
that a further thrusting down would no longer produce
animal-like creatures, but such as were no longer capable
of living. That which had been thrust down into the
animal world has either become extinct or survives in the
different higher animals. Therefore, one must consider
these animals as beings which had to stop at an earlier

stage of human development. They have not retained the form which they had at the time of their separation, however, but have gone from a higher to a lower level. Thus the apes are men of a past epoch who have regressed. As man was once less perfect than he is at present, they were once more perfect than they are now.

That which has remained in the field of the human, has gone through a similar process, but within these human limits. Many savage tribes must be considered to be the degenerated descendants of human forms which were once more highly developed. They did not sink to the level of animalism, but only to that of savagery.

The immortal part of man is the spirit. It has been shown when the spirit entered the body. Before this, the spirit belonged to other regions. It could only associate itself with the body when the latter had attained a certain level of development. Only when one understands completely how this association came about, can one recognize the significance of birth and death, and can understand the nature of the eternal spirit.

viii THE HYPERBOREAN AND
THE POLAREAN EPOCH

T HE FOLLOWING passages from the Akasha Chronicle go back to the periods which precede what was described in the last chapters. In view of the materialistic ideas of our time, the risk we undertake with *these* communications is perhaps even greater than that connected with what has been described in the preceding passages. Today such things are readily met with the accusation of fantasy and baseless speculation. When one knows how far from even taking these things seriously someone can be who has been trained scientifically in the contemporary sense, then only the consciousness that one is reporting faithfully in accordance with spiritual experience can lead one to write about them. Nothing is said here which has not been carefully examined with the means provided by the science of the spirit. The scientist need only be as tolerant toward the science of the spirit as the latter is toward the scientific way of thinking. [Compare my *Welt- und Lebensanschauungen im neunzehnten Jahrhundert* (Conceptions of the World and of Life in the Nineteenth

Century) , where I think I have shown that I am able to appreciate the materialistic-scientific view.*] For those however who incline toward these matters of the science of the spirit, I would like to make a special remark concerning the passages reproduced here. Especially important matters will be discussed in what follows. And all this belongs to periods which are long past. The deciphering of the Akasha Chronicle is not exactly easy in this area. The author of this present book in no way claims that he should be believed blindly. He merely wishes to report what his best efforts have enabled him to discover. He will welcome any correction based on competent knowledge. He feels obliged to communicate these events concerning the development of mankind because the signs of the times urge it. Moreover, a long period of time had to be described in outline here in order to afford a general view. Further details on much that is only indicated now will follow later.

Only with difficulty can the writings in the Akasha Chronicle be translated into our colloquial language. They are more easily communicated in the symbolical sign language used in mystery schools, but as yet the communication of this language is not permitted. Therefore the reader is requested to bear with much that is dark and difficult to comprehend, and to struggle to-

---

* In 1914 a new edition of this work appeared, which was completed by a *Vorgeschichte über abendländische Philosophie und bis zur Gegenwart fortgesetzt* (Previous History of Occidental Philosophy and its Continuation to the Present), the work appearing under the title, *Die Rätsel der Philosophie in ihrer Geschichte als Umriss dargestellt* (The Riddles of Philosophy, etc.), two volumes, Stuttgart, 1955.

ward an understanding, just as the writer has struggled toward a generally understandable manner of presentation. Many a difficulty in reading will be rewarded when one looks upon the deep mysteries, the important human enigmas which are indicated. A true self-knowledge of man is, after all, the result of these "Akasha Records," which for the scientist of the spirit are realities as certain as are mountain ranges and rivers for the eye of sense. An error of perception is of course possible, here and there.

It should be noted that in the present section only the development of man is discussed. Parallel to it, of course, runs that of the other natural realms, of the mineral, the botanical, the animal. The next sections will deal with these. Then much will be spoken of which will make the discussion concerning man appear in a clearer light. On the other hand, one cannot speak of the development of the other terrestrial realms in the sense of the science of the spirit, until the gradual progress of man has been described.

*

If one traces the development of the earth even further back than was done in the preceding essays, one comes upon increasingly refined material conditions of our planet. The substances which later became solid were previously in a fluid, still earlier, in a vaporous and steam-like, and in an even more remote past, in the most refined (etheric) condition. The decreasing temperature caused the hardening of substances. Here we shall go back to the most refined etheric condition of the substances of our earthly dwelling place. Man first entered upon the earth in this epoch of its development. Before that, he belonged to other worlds, which will be discussed later. Only the one immediately preceding will be indicated here. This was a so-called astral or soul world. The beings of this world did not lead an external, (physical) bodily existence. Neither did man. He had already developed the image consciousness mentioned in the previous essay. He had feelings and desires. But all this was enclosed in a soul body. Only to the clairvoyant eye would such a man have been perceptible.

As a matter of fact, all the more highly developed human beings of that time possessed clairvoyance, although it was quite dull and dreamlike. It was not a self-conscious clairvoyance.

These astral beings are in a certain sense the ancestors of man. What is today called "man" carries the self-conscious *spirit* within him. This spirit united with the being which had developed from the astral ancestor in about the middle of the Lemurian period. This union has already been indicated in the previous essays. In the description of the course of development of the ancestors of man up to that period which is to follow here, the matter will be discussed again in greater detail.

The soul or astral ancestors of man were transported to the refined or etheric earth. So to speak, they sucked the refined substance into themselves like a sponge, to speak coarsely. By thus becoming penetrated with sub-

stance, they developed etheric bodies. These had an elon-
gated elliptical form, in which the limbs and other organs
which were to be formed later were already indicated by
delicate shadings of the substance. All processes in this
mass were purely physical-chemical, but they were regulated and dominated by the soul.

When such a mass of substance had attained a certain
size it split into two masses, each of which was similar
to the form from which it had sprung, and in it the same
processes took place as in the original mass of substance.
Each new form was as much endowed with soul as the
mother being. This was due to the fact that it was not
a certain number of human souls which entered upon
the earthly scene, but rather a kind of soul tree which
could produce innumerable single souls from its common root. As a plant sprouts ever anew from innumerable seeds, so the soul life appeared in the countless
shoots produced by the continual divisions. It is true that
from the beginning there was a narrowly circumscribed
number of *kinds* of souls, of which fact we shall speak
later. But within these kinds the development proceded
in the manner which we have described. Each kind of
soul put forth innumerable offshoots.

With their entry into terrestrial materiality, an important change had taken place within the souls themselves.
As long as the souls were not connected with anything
material, no external material process could act on them.
Any action upon them was purely of the nature of soul,
was a clairvoyant one. They thus shared in the life of
everything pertaining to soul in their environment. All

that existed at that time was experienced in this way. The actions of stones, plants, and animals, which then existed only as astral (soul-like) forms, were felt as inner soul experiences.

With the entering upon the earth, something totally new was added to this. External material processes exercised an effect on the soul, which now appeared in material garb. At first it was only the processes of motion in this material outside world which produced movements within the etheric body. As today we perceive the vibration of the air as sound, these etheric beings perceived the vibrations of the etheric matter which surrounded them. Such a being was basically a single organ of hearing. This sense developed first. But one can see from this that separate organs of hearing developed only later.

With the increasing densification of terrestrial matter, the spiritual being gradually lost the ability to mold this matter. Only the bodies which had already been formed could produce their like out of themselves. A new manner of reproduction arose. The daughter being appeared as a considerably smaller form than the mother being and only gradually grew to the size of the latter. While before there had been no organs of reproduction, these now made their appearance.

At this time, however, it is no longer merely a physical-chemical process which takes place in these forms. Such a chemical-physical process could not effect reproduction now. Because of its densification, external matter is no longer such that the soul can give life to it without mediation. Therefore, a certain portion within the form is isolated. This portion is withdrawn from the immediate influences of external matter. Only the body outside of this isolated portion remains exposed to these influences. It is in the same condition in which the whole body was before. In the separated portion, the soul element continues to act. Here the soul becomes the carrier of the life principle, called *Prana* in theosophical literature. Thus the bodily ancestor of man now appears endowed with two organs. One is the physical body, the physical envelope. It is subject to the chemical and physical laws of the surrounding world. The other is the sum of the organs which are subject to the special life principle.

A portion of soul activity is freed in this manner. This activity no longer has any power over the physical part of the body. This part of the soul activity now turns inward and forms a portion of the body into special organs. With this an inner life of the body begins. The body no longer merely participates in the vibrations of the outside world, but begins to *perceive* them within itself as special experiences. Here is the starting point of perception. This perception at first appears as a kind of sense of touch. The organism *feels* the movements of the outside world, the pressure which substances exercise, and so forth. The beginnings of a perception of heat and cold also appear.

With this an important stage in the development of mankind is reached. The immediate influence of the soul has been withdrawn from the physical body. The latter is totally given over to the physical and chemical world of matter. It disintegrates at the moment the soul can

no longer dominate it with its activity. Thereupon occurs
that which one calls "death." In connection with the
preceding conditions, there could be no question of
death. When a division took place, the mother form survived wholly in the daughter forms. For in these the entire transformed soul energy acted as it did before in the
mother form. In the division there was nothing left
which did not contain soul. Now this becomes different.
As soon as the soul no longer has any power over the
physical body, the latter becomes subject to the chemical
and physical laws of the outside world, that is, it dies
away. As activity of the soul there remains only that
which acts in reproduction and in the developed inner
life. This means that descendants are produced by the
force of reproduction, and at the same time these descendants are endowed with a surplus of organ-forming
energy. In this surplus the soul being is constantly re-
viving. As previously at the time of division, the whole
body was filled with soul activity, so the organs of reproduction and perception are now filled with it. We are
thus dealing with a *reincarnation* of the soul life in the
newly-developing daughter organism.

First 2 root races on earth

In theosophical literature these two stages of the development of man are described as the first two root
races of our earth. The first is called the Polarean, the
second, the Hyperborean race.

One must imagine the perceptual world of these ancestors of man to have been a quite general and indefinite
one. Only two of the types of perception of today had
already become separated: the sense of hearing and the

1st

2nd

sense of touch. Because of the changes that had taken place in the body as well as in the physical environment, the entire human form was no longer capable of being, in a manner of speaking, an "ear." A special part of the body remained capable of reverberating to delicate vibrations. It furnished the material from which *our* organ of hearing gradually developed. However, approximately the whole remainder of the body continued to be the organ of touch.

It can be seen that up to this point the entire process of the development of man is connected with a change in the temperature conditions of earth. It was the heat in man's environment which had brought him to the level we have described. But now the external temperature had reached a point where further progress of the human form would no longer have been possible. Within the organism a counter-action against the further cooling of the earth now begins. Man starts to produce his own source of heat. Up to this point he had shared the temperature of his environment. Now organs develop in him which make him able to create the degree of heat necessary for his life. Previously, the circulating substances which passed through him had been dependent on the environment in this respect. Now he himself could develop heat for these substances. The bodily fluids now became warm blood. With this he attained a much higher degree of independence as a physical being than he had possessed before. The whole inner life became more active. Perception still depended entirely on the influences of the outside world. Filled with its own heat, the body acquired an independent *physical* inner life. Now the soul had a basis inside the body upon which it could develop a life which was no longer merely a participation in the life of the outside world.

Through this process, the life of the soul was drawn into the realm of the earthly-material. Previously, desires, wishes, passions, joy and grief of the soul could only be produced by something that was itself soul-like. That which proceeded from another soul-being awakened sympathy or aversion in the soul, excited the passions, and so forth. No external physical object could have had such an effect. Now only did it become possible for such external objects to have a significance for the soul. For the latter experienced the enhancement of the inner life, which had awakened when the body produced its own heat, as something pleasant, the disturbance of this inner life as something disagreeable. An external object suitable for contributing to physical well-being could be *desired*, could be *wished for*. What in theosophical literature is called *Kama*—the body of wishes—became connected with earthly man. The objects of the senses could now become objects of desire. Through his body of wishes man became tied to earthly existence.

This fact coincides with a great event in the universe, with which it is causally connected. Up to this point there had been no material separation between sun, earth, and moon. In their effect on man these three were *one* body. Now the separation took place; the more delicate substantiality, which includes everything which had previously made it possible for the soul to act in an immediately vitalizing manner, separated itself as the sun; the coarsest part was extruded as the moon; and the earth, with respect to its substantiality, stood in the middle between the two others. This separation was of course not a sudden one; rather the whole process proceded gradually while man was advancing from the stage of reproduction by division to the one described last. It was indeed by the universal processes just mentioned that this development of man was brought about. The sun first withdrew its substance from the common heavenly body. Thereby it became impossible for the soul element to vitalize the remaining earthly matter without mediation. Then the moon began to form itself. Thus the earth entered the condition which made possible the capacity for perception characterized above.

In association with this process, a new sense developed. The temperature conditions of earth became such that bodies gradually took on the fixed limits which separated the transparent from the opaque. The sun, which had been extruded from the terrestrial mass, received its task as light giver. In the human body the sense of seeing developed. At first this seeing was not as we know it today. Light and darkness acted upon man as vague sensations. For instance, under certain conditions he experienced light as pleasant, as promoting his physical life, and sought it, strove toward it. At the same time his soul life proper still continued in dreamlike pictures. In this life, colored images which had no immediate relation to external objects arose and vanished. Man still related these colored images to spiritual influences. Light images appeared to him when he was affected by pleasant soul influences, dark images when he was touched by unpleasant soul influences.

Up to now, what was caused by the development of bodily heat has been described as "inner life." But it can be seen that it was not an inner life in the sense of the later development of mankind. Everything proceeds by stages, including the development of the inner life. As it was meant in the preceding essay, this true inner life begins only with the fertilization by the spirit, when man begins to think about that which acts upon him from the outside.

Everything which has been described here shows how man grew into the condition pictured in the preceding chapter. Essentially one is already moving in the period which was characterized there when one describes the following: The soul learns more and more to apply to external bodily existence that which it had previously experienced within itself and related only to the soul-like. This now happens with the colored images. As before, a pleasing impression of something soul-like had been connected with a luminous image in the soul, now a bright impression of light from the outside became connected with such an image. The soul began to see the objects around it in colors. This was connected with the development of new instruments of sight. At previous stages, for the perception of light and darkness, the body had had an eye which no longer exists today. (The legend of the Cyclops with one eye is a recollection of these conditions.) The two eyes developed when the soul began to connect

the light impressions from the outside more intimately
with its own life. With this, the capacity for the perception of the soul-like in the environment disappeared. More
and more the soul became the mirror of the external
world. The outside world is repeated within the soul as
image.

Hand in hand with this went the division into sexes.
On one side, the human body became receptive only to
fertilization by another human being; on the other side
there developed the physical "soul organs" (the nervous
system) through which the sense impressions of the outside world were mirrored in the soul.

With this, the entry of the thinking spirit into the human body had been prepared.

Thought was involved

ix BEGINNING OF THE PRESENT EARTH. EXTRUSION OF THE SUN

WE SHALL now follow the Akasha Chronicle back into that remote past in which our present earth had its beginning. By "earth" is to be understood that condition of our planet by virtue of which it can support minerals, plants, animals, and men in their form of today. For this condition was preceded by others in which the natural realms just mentioned existed in considerably different forms. That which one now calls earth went through many changes before it could become the carrier of our present mineral, plant, animal, and human worlds. Minerals for instance also existed under the preceding conditions, but they looked quite different from those of today. These past conditions will be discussed further below. Now we shall only call attention to the manner in which the immediately preceding condition changed into the present one.

One can conceive of such a transformation to some extent by comparing it to the passage of a plant through the seed stage. Imagine a plant with root, stalk, leaves, blossom, and fruit. It takes in substances from its environment and secretes others. But everything in it which is substance, form, and process disappears, except for the small seed. Life develops by passing through it, and in the new year it rises again in the same form. Thus everything which existed on our earth in its preceding condition has disappeared, only to arise again in its present condition. What for the preceding condition one might call mineral, plant, animal has passed away, as in the plant, root, stalk, and so forth, pass away. There as well as here, a germinal stage has remained, from which the old form develops anew. The forces which will cause the new form to emerge lie hidden in the seed.

At the period discussed here we are dealing with a kind of earth germ. This contained within itself the forces which led to the earth of today. These forces were acquired through earlier conditions. This earth germ however must *not* be imagined as a densely material one, like that of a plant. It was rather of a soul character. It consisted of that delicate, malleable, mobile substance which is called "astral" in occult literature.

In this astral germ of earth there are *only* human rudiments at first. These are the rudiments of the later human souls. Everything in preceding conditions which was already present as a mineral, plant, or animal nature has been drawn into these human rudiments and become fused with them. Before man enters upon the earth he is a *soul*, an astral entity. As such he appears on earth. The latter exists in a state of the most highly-refined substantiality, which in occult literature is called the *most refined*

ether. Whence this etheric earth originated will be described in the next essays.

The astral human beings combine with this ether. They
impress their nature upon this ether, in order that it can
become a likeness of the astral human entity. In this
initial condition we are dealing with an ether earth which
really consists only of these ether men, which is only a
conglomerate of them. Actually the astral body or the
soul of man is for the most part still *outside* the ether
body and organizes it from without. To the scientist of
the spirit, the earth appears approximately as follows. It
is a sphere which in turn is composed of innumerable
small ether spheres—the ether men—and is surrounded by
an *astral* envelope just as the present earth is surrounded
by an envelope of air. It is in this astral envelope (atmosphere) that the astral men live and whence they act upon
their ether likenesses. The astral human souls create organs in their ether likenesses and produce a human ether
life in them. Within the whole earth there exists only
one condition of matter, the refined living ether. In theosophical books this first humanity is called the first (the
Polarean) root race.

The further development of earth takes place in such
a way that from the *one* condition of matter there develop
two. A denser substantiality is secreted, so to speak, and
leaves a thinner one behind. The denser substantiality
resembles our present air; the thinner one is that which
causes chemical elements to develop from previously un-
differentiated substance. Along with these, a remainder
of the previous substantiality, the living ether, continues

1— Denser substantiality
   resembles our present air—
2— The thinner one
3— a remainder of previous substantiality
   continues to exist.

to exist. Only a part of it is transformed into the so-called
material conditions. We now are dealing with three substances within the physical earth. While the astral human
beings in the envelope of earth previously acted only upon
one kind of substantiality, they must now act upon three.
They act upon them in the following way: That which
has become airlike at first resists their activity. It does not
accept everything which is rudimentarily present in the
complete astral men. As a consequence, astral humanity
must divide itself into two groups. One group works on
the airlike substantiality and creates in it a likeness of
itself. The other group can do more. It can work on the
two other substantialities; it can create a likeness of itself
which consists of the living ether *and* of the other kind
of ether which brings the elementary chemical substances
into being. This ether will here be called the chemical
ether. This second group of astral men has acquired its
higher capacity, however, only by separating from itself
a part of the astral nature—the first group—and condemn-
ing it to a lower kind of labor. Had it retained within
itself the forces which accomplish this lower labor, it
could not have risen higher itself. Therefore, here we are
dealing with a process which consists in the development
of the higher at the expense of something else, which is
separated from it.

Within the physical earth the following picture now
presents itself. Two kinds of entities have come into being.
First, entities which have an airlike body on which the
astral being belonging to it is working from the outside.
These beings are animal-like. They form a first animal

realm on earth. These animals have shapes which, were
they to be described here, would strike mankind of today
as very peculiar. Their shape—one must keep in mind
that this shape is based only on an airlike substance—does
not resemble any of the animal forms existing now. At
most they have a remote similarity to the shells of certain
snails and mussels which exist today. Beside these animal
forms the development of physical man progresses. The
astral man, who has now ascended higher, produces a physical likeness of himself which consists of the two kinds of
matter, of the life ether and of the chemical ether. One
thus deals with a man who consists of the astral body and
is working himself into an ether body which in turn consists of two kinds of ether: life ether and chemical ether.
Through the life ether this physical likeness of man is
endowed with the capacity to reproduce itself, to cause
beings of its own kind to emerge from it. Through the
chemical ether it develops certain forces which are similar
to the present forces of chemical attraction and repulsion.
Thereby this likeness of man is in a position to attract
certain substances from the environment and to combine
them with itself, secreting them again later by means of
the repelling forces. These substances, of course, can only
be taken from the animal realm described above, and
from the realm of man. This constitutes a beginning of
nutrition. Thus these first likenesses of man were eaters
of animals and of men.

Besides these beings, the descendants of the earlier beings, composed merely of life ether, continue to exist, but

they become atrophied, since they have to adapt to the new terrestrial conditions. After they have undergone many transformations, the unicellular animal beings develop from them, and also the cells which later make up the more complicated living organisms.

The following process then takes place. The airlike substantiality divides itself into two, of which one becomes denser, watery, while the other one remains airlike. The chemical ether also divides itself into two conditions of matter; one of them becomes denser and forms that which we shall here call the light ether. It endows the entities which possess it with the gift of luminosity. On the other hand, a portion of the chemical ether continues to exist as such.

We are now dealing with a physical earth which is composed of the following kinds of matter: water, air, light ether, chemical ether, and life ether. In order that the astral entities can act on these kinds of matter, another process takes place by which the higher develops at the expense of the lower, which becomes separated from it. Thereby physical entities of the following kind are produced. First, those whose physical body consists of water and air. Now coarse astral entities which have been split off, act on these. Thus a new group of animals of coarser materiality than the earlier ones is produced.

Another new group of physical entities has a body which consists of air and light ether mixed with water. These are plantlike entities, which however are very different in form from the plants of today. Finally, the third new group represents man of that period. His physical body consists of three kinds of ether: the light ether, the chemical ether, and the life ether. If one considers that descendants of the old groups also continue to exist, one can judge what a variety of living beings there already were at that stage of terrestrial existence.

There now follows an important cosmic event. The sun is extruded. Thereby certain forces simply leave the earth. These forces are composed of a part of what hitherto had existed on earth in the life ether and in the chemical and light ether. These forces, so to speak, were withdrawn from the earth. A radical change thereby took place among all groups of terrestrial beings which previously had contained these forces within themselves. They all suffered a transformation. Those which have been called plant beings above, first suffered such a transformation. A part of their light ether forces was taken from them. They could then develop as organisms only when the force of light, which had been withdrawn from them, acted upon them from the outside. Thus the plants came under the influence of the sunlight.

Something similar happened with human bodies. From then onward, their light ether also had to act together with the light ether of the sun in order to be capable of life. But not only those beings themselves which lost the light ether were affected; the others were affected too. For in the world everything interacts. Those animal forms, too, which did not contain light ether themselves had previously been irradiated by their fellow beings on earth and had developed under this irradiation. Now they also came under the immediate influence of the external sun.

The human body in particular developed organs receptive to the sunlight, that is, the first rudiments of human eyes.

The consequence of the extrusion of the sun was a further material densification of the earth. Solid matter developed from fluid; likewise the light ether separated into another kind of light ether, and into an ether which gives bodies the capacity to increase temperature. With this, the earth became an entity which developed heat within itself. All its beings came under the influence of heat. In the astral element a process similar to the previous ones again had to take place; some beings developed to a higher level at the expense of others. A group of beings split off which were well suited to work on coarse, solid substantiality. With this there had developed the firm skeleton of the *mineral realm* of earth. At first the higher natural realms did not act upon this rigid mineral skeleton. Thus, on the earth there exist a mineral realm which is solid, and a plant realm which has water and air as its densest substantiality. In the latter realm, through the events we have described, the air body had become condensed to a water body. There also existed animals of the most varied forms, some with water and some with air bodies. The human body itself had become subject to a process of densification. It had condensed its most compact corporeality to the point of wateriness. The newly-develped heat ether coursed through this water body. This gave to his body a substantiality which could perhaps be called gaslike. This material condition of the human body is described in works on mystery science as that of the *fire mist.* Man was embodied in this body of fire mist.

With this, the examination of the Akasha Chronicle has reached a point shortly before the cosmic catastrophe caused by the extrusion of the moon from the earth.

✶ EXTRUSION OF THE MOON

Oₙₑ must be quite clear about the fact that only later did man assume the dense substantiality which he has today, and that he did this very gradually. If one wants to form an idea of his corporeality on the level of development which is being discussed, one can best do this by imagining it as similar to water vapor or to a cloud suspended in the air. But of course this idea approaches reality in a completely *external* way. For the fire cloud "man" is internally alive and organized. In comparison to what man became later, however, one must imagine him at this stage as in a state of soul-slumber, and as only very dimly conscious. Everything which can be called intelligence, understanding, reason is lacking in this being. Floating rather than striding, it moves forward, aside, backward, to all sides, by means of four limblike organs. For the rest, something has already been said about the soul of this being.

One must not think however that the movements or vital activities of these beings occurred in an irrational or irregular fashion. On the contrary, they were com-

pletely regular. Everything which happened had sense and
significance. But the directing force of understanding was
not in the beings themselves. They were directed by an
understanding which was outside of them. Higher, more
mature beings than they, surrounded and directed them.
For the important, basic quality of the fire mist was that
on the level of their existence which we have characterized,
human beings could embody themselves in it, but that
at the same time higher beings also could take on a body
in it and could enter into a fully reciprocal relationship
with men. Man had brought his impulses, instincts, and
passions to the point where they could be formed in the
fire mist. The other beings mentioned, however, could
create within this fire mist by means of their reason and
their intelligent activity. These beings had higher capacities by which they reached into the upper regions.
Their decisions and impulses emanated from these regions,
but the actual effects of these decisions appeared in the
fire mist. Everything men did on earth resulted from the
regular association of the fire mist body with that of these
higher beings.

One can say that man was striving to ascend. He was to
develop qualities in the fire mist which in a human sense
were higher than those he had previously possessed. The
other beings, however, were striving downward toward
the material. They were on the way to bringing their
creative powers to bear on increasingly dense material
forms. This does not represent a degradation for them in
the broader sense of the term. One must be quite clear on
this point. It requires a higher power and capacity to

direct denser forms of substantiality than to control those less dense. In earlier periods of their development, these higher beings too had had a limited power like that of man today. Like present-day man, they once had power only over what took place "within them." At that time, coarse, external matter did not obey them. Now they were striving toward a condition in which they were to direct outer events magically. Thus they were ahead of man in the period described. Man strove upward in order that he might first embody the understanding in more refined matter, so that later it could act toward the external; they had already incorporated the understanding into themselves at an earlier period, and now received magic power in order to articulate the understanding into the world around them. Man was moving *upward* through the stage of the fire mist; they were penetrating *downward* through the same stage, toward an extension of their power.

Those forces especially, which man knows as the forces of his lower passions or impulses, can be active in the fire mist. Man, as well as the higher beings, makes use of these forces at the stage of the fire mist. These forces act in such a way within the human form described above that man can develop the organs which enable him to think, and thus to develop a personality. On the other hand, these forces work in the higher beings at this stage in such a manner that they can employ them impersonally to create the arrangements of the earth. In this way, forms which are images of the rules of the understanding, come into existence on earth through these beings. Through the action of the forces of passion, organs of personal understanding develop in man; through the same forces, organizations filled with wisdom develop around him.

One should now imagine this process to be somewhat further advanced; or rather, one should represent to oneself what is written in the Akasha Chronicle concerning a somewhat later point in time. At that moment the moon split off from the earth. This event caused a great revolution. The objects which surround man lost a great part of their heat. These objects thereby entered into a coarser and denser substantiality. Man must live in this cooler environment. He can do this only if he changes his own substantiality. With this densification of substance is connected a change in form. For the condition of fire mist on earth has been replaced by a completely different state. As a consequence, the higher beings which we have described no longer have the fire mist available to them as a medium for their activity. Now they can no longer exercise their influence on those soul activities of man which had previously constituted their main field of action. They have received power over the forms of man which they themselves had previously created from the fire mist.

This change in influence goes hand in hand with a transformation of the human form. One half of this form, together with two organs of movement, now becomes the lower half of the body, which functions mainly as the carrier of nutrition and reproduction. The other half of this form is turned upward, so to speak. The remaining two organs of movement become the

rudiments of hands. Those organs which previously had
served for nutrition and reproduction are transformed
into organs of speech and thought. Man has become up-
right. This is the immediate consequence of the extrusion of the moon. With the moon all those forces
disappeared from the earth through which, during his
fire mist period, man could still impregnate himself and
produce beings like himself without external influence.
His whole lower half—that which one often calls the
lower nature—now came under the rationally formative
influence of the higher entities. What these entities previously could regulate within man, since the mass of
forces now split off with the moon was then still combined with the earth, they now have to organize through
the interaction of the two sexes. It is therefore understandable that the moon is regarded by the initiates as
the symbol of the force of reproduction. After all, these
forces do inhere in it, so to speak. The higher beings
we have described have an affinity with the moon, are in
a sense, moon gods. Before the separation of the moon
and, through its power, they acted within man; after-
wards, their forces acted from outside on the reproduction of man. One could also say that those noble
spiritual forces which previously had acted on the still
higher impulses of man through the medium of the
fire mist, had now descended in order to exercise their
power in the area of reproduction. Indeed, noble and
divine forces exercise a regulating and organizing action
in this area.

With this an important proposition of the secret doctrine has been expressed, namely, the higher, more noble divine forces have an affinity with the—*apparently*—lower forces of human nature. The word "apparently" must here be understood in its full significance. For it would be a complete misconception of occult truths if one were to see something base in the forces of reproduction as such. Only when man misuses these forces, when he compels them to serve his passions and instincts, is there something pernicious in them, but not when he *ennobles* them through the insight that a divine spiritual power lies in them. Then he will place these forces at the service of the development of the earth, and through his forces of reproduction he will carry out the intentions of the higher entities we have characterized. Mystery science teaches that this whole subject is to be ennobled, is to be placed under divine laws, but is not to be mortified. The latter can only be the consequence of occult principles which have been understood in a purely external fashion and distorted into a misconceived asceticism.

It will be seen that in his second, his upper half, man has developed something upon which the higher beings we have described have no influence. Other beings now acquire power over this upper half. In earlier stages of their development, these beings advanced further than men, but not as far as the moon gods. They could not exercise their power in the fire mist. But now that something they themselves had lacked previously has been formed in the human organs of understanding through the fire mist, their time has come. At an earlier time, the moon gods had attained an understanding capable of acting externally. This understanding already existed in them when the period of the fire mist began. They could act externally on the things of earth. In earlier times, the lower beings we have just mentioned had not attained such an understanding which acts outwardly. Therefore, the time of the fire mist found them unprepared. Now, however, an understanding is present. It exists in men. These beings seize upon this human understanding in order to act on the things of earth by means of it. As the moon gods previously had acted on the *whole* man, they now act only on his lower half, while the influence of the lower entities just mentioned acts on his upper half. Thus man comes under a double leadership. In his lower part he is under the power of the moon gods; in his developed personality, however, he comes under the leadership of those entities which are summed up under the name "Lucifer," the name of their regent. The Luciferic gods thus complete their own development by making use of the awakened human powers of understanding. Previously they had not been able to attain this level. At the same time they give man the predisposition to freedom, to the discrimination between "good" and "evil." While it is true that the human organ of understanding has been formed entirely under the leadership of the moon gods, these gods would have left it to slumber; they were not interested in making use of it. They possessed their own powers of understanding. In their own interest, the Luciferic beings were concerned with developing the human understanding and

directing it toward the things of earth. Thereby for men
they became the teachers of all that can be accomplished
by the human understanding. But they could not be
anything more than *stimulators.* They could not develop
an understanding *within themselves,* but only *in man.*
Thus there developed two directions of activity on earth.
One proceeded directly from the moon divinities and was
lawfully regulated and rational from the very beginning.
The moon gods had already served their apprenticeship
and were now beyond the possibility of error. The Luciferic gods which acted on men had yet to win their way
to such illumination. With their guidance man had to
learn to find the laws of his being. Under Lucifer's leadership he himself had to become as "one of the gods."

Here the question arises: If in their development the
Luciferic entities had not reached the stage of intelligent
creation in the fire mist, at what stage had they stopped?
To what point in earth development were they able to
work together with the moon gods? The Akasha Chronicle
gives information on this. They could participate in
earthly creation up to the point at which the *sun* split
off from the earth. It appears that while they performed
somewhat lesser work than the moon gods up to this time,
nevertheless they belonged to the host of divine creators.
After the separation of earth and sun, an activity began
on the earth—the work in the fire mist—for which only
the moon gods, but not the Luciferic spirits, were prepared. Therefore, a period of pause and of waiting began
for these spirits. The Luciferic spirits could emerge once
more from their state of rest when the human beings be-

gan to work at the development of their organs of understanding, after the ebbing of the general fire mist. For
the creation of the understanding is related to the activity of the sun. The dawn of the understanding in human nature is the lighting up of an inner sun. This is
said not only in a metaphorical, but also in a quite real
sense. When the epoch of the fire mist had ebbed from
the earth, these spirits found within man an opportunity
to resume their activity connected with the *sun*.

It now becomes clear whence the name Lucifer, that
is, "the bearer of light," originates, and why these beings
are designated as "sun gods" in mystery science.

All that follows can only be understood if one looks
back to periods preceding the development of the earth.
This will be done in the next chapters of the "Akasha
Chronicle." The development through which the beings
connected with earth passed on other planets before appearing on the earth, will be shown there. In addition,
one will become more fully acquainted with the nature
of the "moon gods" and of the "sun gods." Simultaneously,
the development of the animal, plant and mineral realms
will become entirely clear.

WE SHALL NEXT consider the development of man and of the entities connected with him in the time which preceded the "earthly period." For when man began to unite his destiny with the planet one calls "earth," he had already passed through a series of developmental steps in the course of which he had prepared himself for earthly existence, as it were. One must distinguish three such steps, which are designated as *three planetary developmental stages*. The names used in mystery science for these stages are the Saturn, Sun and Moon periods. It will become apparent that these designations *at first* have nothing to do with the heavenly bodies of today which bear these names in physical astronomy, although in a *broader* sense a relationship to them exists, which is known to the advanced mystic.

One will sometimes say that man inhabited other planets before he appeared upon earth. But under these "other planets" one must only understand earlier developmental conditions of the earth itself and of its inhabitants. Before it became "earth," the earth with all the

beings which belong to it passed through the three conditions of the Saturn, Sun, and Moon existence. Saturn,
Sun, and Moon are, as it were, the three incarnations of
the earth in primeval times. What in this connection is
called Saturn, Sun, and Moon no more exists today as a
physical planet than the previous physical incarnations
of a human being continue to exist alongside his present
one.

This "planetary development" of man and of the other
beings belonging to earth will form the subject of the
following discussions "From the Akasha Chronicle." By
this we do not wish to say that the three conditions named
were not preceded by others. But everything which precedes these three is lost in a darkness which for the
present the research of mystery science cannot illuminate.
For this research is not based on speculation, on a day-
dreaming in terms of mere concepts, but on actual *spiritual experience*. As our physical eye can see outdoors
only as far as a certain boundary line and cannot look
beyond the horizon, so the "spiritual eye" can look only
as far as a certain point in time. *Mystery science is based
on experience and is content to remain within this experience*. Only in a conceptual splitting of hairs will one
want to find out what was "at the very beginning" of
the world, or "why God really created the world." For
the scientist of the spirit it is rather a matter of realizing
that at a certain stage of cognition one no longer poses
such questions. Everything man needs for the fulfillment
of his destiny on our planet is revealed to him within
spiritual *experience*. The one who patiently works his

way into the experiences of scientists of the spirit will see that *within* spiritual experience man can obtain full satisfaction concerning all those questions which are vital to him. In the following essays for example, one will see how completely the question concerning the "origin of evil" is resolved, as well as much else which man must desire to know.

We by no means intend to imply that man can *never* receive enlightenment concerning questions about the "origin of the world" and similar matters. *He can*. But in order to be able to be enlightened, he must first absorb the knowledge revealed *within* more proximate spiritual experience. He then comes to realize that he must ask these questions in a different manner than heretofore.

The more deeply one works his way into true mystery science, the more *modest* he becomes. Only then does he realize how one must very gradually make oneself ready and worthy for certain insight. Pride and arrogance finally become names for human qualities which no longer make sense at a certain level of cognition. When one has understood a little, he sees how immeasurably long is the road which lies ahead of him. Through knowledge one gains insight into "how little one knows." He also acquires a feeling for the immense responsibility he assumes when he speaks of supersensible cognition. But mankind cannot live without the latter. However, he who promulgates such knowledge needs modesty and true self-criticism, an unshakable striving for self-knowledge and the utmost caution.

Such remarks are necessary here, since now the ascent

toward even higher knowledge than is to be found in the preceding sections of the "Akasha Chronicle," is to be undertaken.

To the vistas which in the following essays will be opened toward the past of man, others will be added *upon the future*. For the future can be revealed to true spiritual cognition, if only to the extent to which this is necessary for man in order that he can fulfill his destiny. The one who will have nothing to do with mystery science and from the judgment-seat of his prejudices, simply consigns everything coming from that quarter to the realm of fantasy and dreams—he will understand this relationship to the future least of all. Yet a simple logical consideration could make clear what is in question here. But such logical considerations are accepted only when they coincide with the preconceptions of men. Prejudices are mighty enemies of logic.

If sulphur, oxygen, and hydrogen are brought together under certain definite conditions, sulphuric acid must be produced, according to an inevitable law. The student of chemistry can *predict* what must happen when these three elements come into contact with one another under given conditions. Thus, such a student of chemistry is a prophet in the limited field of the material world. His prophecy could only prove false if the laws of nature were suddenly to change. Now the scientist of the spirit investigates spiritual laws in just the manner in which the physicist or the chemist investigates material laws. He does this in the manner and with the exactness which are requisite in the spiritual field. However, the development of mankind depends on these great spiritual laws. Just as little as oxygen, hydrogen, and sulphur will combine at some future time in a manner contrary to laws of nature, so little will anything occur in the spiritual life which is contrary to spiritual laws. The one who knows these spiritual laws can look into the *orderliness of the future*.

The use of precisely this comparison for the prophetic prediction of the coming destinies of mankind is intentional here, because true mystery science really understands this prediction in just this sense. For the one who forms a clear idea of this conviction of occultism, the objection that any human freedom is made impossible because events can be predicted in a certain sense, becomes void. That can be predicted which is in accordance with a law. But the will is *not* determined by a law. Just as it is certain that in *each* case oxygen, hydrogen, and sulphur are combined into sulphuric acid only according to a definite law, just so is it equally certain that the establishing of the conditions under which the law will act, can depend on the human will. Thus it will be with the great world events and human destinies of the future. As a scientist of the spirit, one foresees them, although they are to be brought about only by human choice. He foresees what is accomplished by the freedom of man. The following essays will show that this is possible.

However, one must be clear about *one* essential difference between the prediction of events through physical science and that through spiritual cognition. Physical science is based on the insights of the understanding, and therefore its prophecy is only based on the intellect, which has to rely on judgments, deductions, combinations, and so forth. *Prophecy through spiritual cognition,* on the contrary, proceeds from an actual *higher seeing* or *perceiving.* The scientist of the spirit must strictly avoid even representing anything to himself which is based on mere reflecting, combining, speculating, and so forth. Here he must practice the most far-reaching renunciation and be quite clear that all speculating, intellectual philosophizing, and so forth is a hindrance to true seeing. These activities still belong entirely to the lower nature of man, and truly higher cognition begins only when this nature raises itself to the higher nature in man. Here nothing is really said against these activities, which are not only wholly justified in their field, but are there the *only* justified ones. In itself, a thing is neither higher nor lower; it is higher or lower only in relation to something else. What is high in one respect can be very low in another.

However, what must be understood through *seeing,* *cannot* be understood through mere reflection or through even the most magnificent combinations of the intellect. A person may be ever so "ingenious" in the usual sense of the word, but this "ingenuity" will avail him absolutely nothing with respect to the cognition of supersensible truths. He must even renounce it, and abandon himself solely to the higher seeing. Then he will perceive things without his "ingenious" reflecting, just as he perceives the flowers in the fields without further reflection. It does not help one to reflect about the appearance of a meadow;

all intellect is powerless there. The same is true of the
seeing into higher worlds.

What can be said prophetically in this way about the
future of man is the basis for all *ideals* which have a real,
practical significance. If they are to have value, ideals
must be rooted as deeply in the spiritual world as are
natural laws in the natural world. Laws of development
must be such true ideals. Otherwise they spring from a
gushing enthusiasm and a fantasy which are valueless,
and can never be fulfilled. In the broadest sense, all
great ideals of world history have proceeded from clear
cognition. For, in the final analysis, all these great ideals
originate with the great scientists of the spirit or initiates,
and those lesser ones who collaborate in the development
of humanity direct themselves either consciously or—most
often—*unconsciously* in accordance with the instructions
of the spiritual scientists. Everything unconscious must
finally have its origin in something conscious. The brick-
layer who works on a house "unconsciously," directs himself according to matters of which others are conscious
who have determined the place where the house is to be
built, the style in which it is to be erected, and so forth.
But this determining of place and style is based on something of which the determiners remain unconscious, but
of which others are or were *conscious.* An artist, for example, knows why a particular style requires a straight
line here, a curved line there, and so forth. The one who
uses this style for his house perhaps does not become
conscious of this "why."

This is also the case with the great events in the development of the world and of mankind. Behind those who work in a certain field stand higher, more conscious workers, and thus the scale of consciousness goes up and down.

Behind the general mass of men stand the inventors, artists, scientists, and so forth. Behind them stand the initiates of mystery science, and behind them stand super-human beings. The development of the world and of mankind becomes comprehensible only if one realizes that ordinary human consciousness is but *one* form of consciousness, and that there are *higher* and *lower* forms. But here too one must not misapply the expressions "higher" and "lower." They have a significance only in relation to the point where one happens to be standing. It is no different with this than with "right and left." When one stands at a certain place, some objects are "right" or "left" of him. If one moves a little to the "right," objects which before were on the right, are now on the left. The same is true of the levels of consciousness which lie "higher" or "lower" than ordinary human consciousness. When man himself develops more highly, his relations to the other levels of consciousness change. But these changes are connected with his development. It is therefore important to indicate such other levels of consciousness here by means of examples.

The beehive or that magnificent commonwealth embodied in an ant hill provide bases for such an indication. The collaboration of the various kinds of insects (females, males, workers) proceeds in a completely systematic fashion. The distribution of tasks among the

several categories can only be described as an expression
of true wisdom. What happens here is just as much the
result of a consciousness as the institutions of man in
the physical world (technology, art, state, and so forth)
are an effect of his consciousness. However, the consciousness at the base of the beehive or the ant society is not
to be found in the same physical world in which the
ordinary human consciousness exists. In order to describe the situation, one can express oneself somewhat as
follows. One finds man in the physical world. His physical organs, his whole structure are such that at first one
looks for his consciousness also in this physical world. It
is otherwise with the beehive or the ant hill. Here it
would be quite wrong to confine oneself to the physical
world with respect to the consciousness in question, as
was done in the case of man. No, here one must say that
to find the ordering principle of the beehive or the ant
hill, one cannot confine oneself to the world where the
bees or ants live in their physical bodies. In this case, the
"conscious mind" must be sought directly in another
world. The same conscious mind which in man lives
in the physical world, in the case of these animal colonies
must be sought in a *supersensible* world. If with his
consciousness man could raise himself into this supersensible world, he would be able to greet the "ant or bee
spirit" there in full consciousness as his sister being. *The
seer can actually do this.* Thus, in the examples given
above, we are confronted by beings which are conscious
in other worlds and which reach into the physical world
only through their physical organs—the individual bees

and ants. It is quite possible that a consciousness like
that of the beehive or of the ant hill existed in the physical world in earlier periods of its development, as that
of man does now, but then raised itself and left behind
in the physical world only its acting organs, that is, the
individual ants and bees. Such a course of development
will actually take place in the future with respect to man.
In a certain manner it has already taken place among the
seers in the present. That the consciousness of contemporary man functions in the physical world is due to the
fact that its physical particles—the molecules of brain
and nerves—exist in a quite definite relation with one
another. What has been discussed in greater detail in
another connection in my book, *Wie erlangt man Erkent-
nisse der höheren Welten?* (How Does One Attain Knowledge of Higher Worlds?) will also be indicated briefly
here. In the course of the higher development of man,
the ordinary connection of the brain molecules is dis-
solved. They are then connected more "loosely," so that
the brain of a seer can really be compared with an ant
hill in a certain respect, although the segmentation is
not demonstrable *anatomically*. In different activities of
the world these processes occur in quite different ways.
At a time long past, the individual molecules of the ant
hill—that is, the ants themselves—were firmly connected,
just as are the molecules of the human brain today. At
that time, the consciousness corresponding to them was
in the physical world, as that of man is today. When
human consciousness will travel into "higher" worlds
in the future, the connection between the material parts

in the physical world will be as loose as is that between the individual ants today. What in time will occur physically in all men, already takes place today in the brain of the clairvoyant, but no instrument of the world of the senses is sufficiently delicate to show the loosening which comes about through this anticipatory development. Just as among the bees three categories, queens, drones, workers, are formed, so three categories of molecules are formed in the "seer brain," molecules which are actually individual, living beings, brought into conscious collaboration by the consciousness of the seer, which is in a higher world.

Another level of consciousness is represented by what one usually calls the *folk- or racial spirit*, without representing anything very definite to oneself by this. For the scientist of the spirit, a consciousness also exists at the base of the common and wise influences which appear in the communal life of the members of a people or of a race. Through occult research, one finds this consciousness to be in another world, just as was the case with the consciousness of a beehive or of an ant hill. However, there are no organs for this "folk" or "racial consciousness" in the physical world; rather these organs are to be found only in the so-called astral world. As the consciousness of the beehive works through the physical bees, so the folk-consciousness works by means of the astral bodies of the human beings belonging to a people. In these "folk- and racial spirits" one is therefore confronted with kinds of entities quite different from those in man or in the beehive. Many more examples would have to be given in order to show clearly how subordinate and superior entities exist in relation to man. But it is hoped that what has been given will be sufficient to introduce the avenues of human development described in the following chapters. For the development of man himself can only be understood when one considers that he develops together with beings whose consciousness exists in other worlds than his own. What happens in his world is also dependent on these beings who are connected with other levels of consciousness, and therefore can be understood only in relation to this fact.

As INDIVIDUAL man has to pass through different stages after his birth, as he must ascend from infancy through childhood and so on to the age of the mature adult, so too must mankind as a whole go through a similar process. Humanity has developed to its present condition by passing through other stages. With the methods of the clairvoyant one can discern three principal stages of this development of mankind which were passed through before the formation of the earth took place and before this sphere became the scene of that development. Therefore at present we are concerned with the fourth stage in the great universal life of man. For now we shall relate the relevant facts here. The deeper explanation will appear in the course of the description, insofar as is possible in the words of ordinary language, that is, without having recourse to the form of expression of mystery science.

Man existed before there was an earth. But one must not imagine—as has already been suggested—that perhaps he had previously lived on other planets and then at a certain time migrated to earth. Rather, the earth has developed together with man. Just as man has passed through three main stages of development, so has the earth, before becoming that which one now calls "earth." For the time being, as has been indicated above, one must completely liberate oneself from the significance which contemporary science connects with the names "Saturn," "Sun," and "Moon," if one wants to see the explanations of the scientist of the spirit in this area in their proper light. For the present one should connect with these names no other significance than that directly given to them in the following communications.

Before the heavenly body on which the life of man takes place became "earth," it had had three other forms, which one designates as Saturn, Sun, and Moon. One can thus speak of four planets on which the four principal stages of the development of mankind take place. It is so that before the earth became "earth," it was Moon, before that Sun, and yet earlier, Saturn. One is justified, as will appear from the following communications, to assume three further principal stages which the earth, or better the heavenly body which developed into the present earth, still has to pass through. In mystery science these have been named Jupiter, Venus, and Vulcan. Thus the heavenly body with which human destiny is connected has passed through three stages in the past, is now in its fourth, and will in the future have to pass through three more until all the talents which man has within himself are developed, until he arrives at the peak of his perfection.

One must realize that the development of man and of his heavenly body does not proceed as gradually as for instance the passage of an individual human being through infancy, adolescence and so forth, where one condition goes over into another more or less imperceptibly. Rather there are certain interruptions. The Saturn condition does not go over immediately into the Sun stage. Between Saturn development and Sun development, and similarly between the subsequent forms of the heavenly body inhabited by man, there are intermediate conditions which can be compared with the night between two days or with the sleeplike condition of a plant seed before it again develops into a full plant.

In imitation of oriental descriptions of this state of affairs, contemporary theosophy calls a stage of development in which life is externally furthered, *Manvantara,* the intermediate condition of rest, *Pralaya.* In accordance with the usage of European mystery science, one can use the word "open cycle" for the former condition, and on the other hand, "hidden or closed cycle" for the latter. But other designations are also in common use. Saturn, Sun, Moon, earth, and so forth, are "open cycles," and the periods of rest between them are "closed" ones.

It would be quite erroneous to think that in the periods of rest all life is extinct, although today this idea can be encountered in many theosophical circles. Just as little as man ceases to live during his sleep, so little does his life and that of his heavenly body become extinct during a "closed cycle" (*Pralaya*). It is only that the conditions of life in the periods of rest cannot be perceived with the senses which have been developed during the "open cycles," just as during his sleep man does not perceive what is taking place around him. Why one uses the expression "cycle" for the stages of development will become sufficiently clear in the course of the following discussion. Only later can we speak about the enormous periods of time which are required for these "cycles."

One can find a thread through the course of the cycles by following for a moment the development of human *consciousness* through them. Everything else can suitably arise out of this consideration of consciousness.

The consciousness which man develops during his life-course on earth will be called—in accordance with European mystery science—the "clear consciousness of day." The latter consists in the fact that through his present senses, man perceives the things and beings of the world and that he forms conceptions and ideas concerning these things and beings with the help of his understanding and of his reason. He then acts in the world of the senses according to these perceptions, conceptions, and ideas. Man formed this consciousness only in the fourth principal stage of his cosmic development; on Saturn, Sun, and Moon it did not yet exist. There he lived in other conditions of consciousness. As a result, one can describe the three previous stages of development as the unfolding of lower conditions of consciousness.

The lowest condition of consciousness was passed

through during the Saturn development; the Sun condition is higher, then follows the Moon consciousness and
finally that of earth.

These former consciousnesses are primarily distinguished from the earthly one by two characteristics: by
the degree of clarity, and by the area over which the
perception of man extends.

The Saturn consciousness has the lowest degree of clarity. It is entirely dull. It is difficult to give an exact idea
of this dullness, since even the dullness of sleep is somewhat clearer than this consciousness. In abnormal, so-
called deep states of trance, modern man can still fall back
into this state of consciousness. The clairvoyant in the
sense of mystery science can also form a correct conception of it. But by no means does he himself live in this
state of consciousness. On the contrary, he ascends to a
much higher one, which however in certain respects is
similar to the original one. In the ordinary man at the
contemporary terrestrial stage, this condition, through
which he once passed, has been effaced by the "clear consciousness of day." The "medium" who falls into a deep
trance, however, is transported back into it, so that he
perceives in the same way in which all men perceived
during the "Saturn period." Either during the trance or
after awaking, such a medium can then tell of experiences which are similar to those of the Saturn stage. One
must be careful to say that they are "similar," not "identical," for the events which took place on Saturn are once
and for all past; only events which have a certain affinity
with them still take place in the environment of man.

These can only be perceived by a "Saturn consciousness."

Like the medium, the clairvoyant in the above sense acquires such a Saturn consciousness, but in addition to it he keeps his "clear consciousness of day," which man did not yet have on Saturn, and which the medium loses in the state of trance. Such a clairvoyant is not in the Saturn consciousness itself, but he can form a conception of it.

While this Saturn consciousness is by some degrees inferior to the one of today with respect to clarity, it is superior to the latter with respect to the extent of what it can perceive. In its dullness it can not only perceive everything which takes place on its own heavenly body down to the last detail, but it can also observe the objects and beings on other heavenly bodies which are connected with Saturn. It can also exercise a certain influence on these objects and beings. (It hardly need be said that this observation of other heavenly bodies is quite different from that which contemporary man can undertake by means of his scientific astronomy. This astronomical observation is based on the "clear consciousness of day" and therefore perceives other heavenly bodies from the outside. The Saturn consciousness, on the other hand, is immediate sensation, an experiencing of what takes place on other heavenly bodies. One does not speak altogether accurately, but still fairly so, if one says that an inhabitant of Saturn experienced objects and events of other heavenly bodies—and of his own—as a man of today experiences his heart and his heartbeat or something similar in his own body.)

This Saturn consciousness developed slowly. As the first principal stage in the development of mankind it passed through a series of subordinate stages, which in European mystery science are called "small cycles." In theosophical literature it has become customary to call these "small cycles," "rounds," and their further sub-divisions—still smaller cycles—"globes." These subordinate cycles will be dealt with in subsequent discussions. For the sake of greater clarity, we shall first follow here the principal stages of development. For the moment we shall speak only of man, although the development of subordi-nate and superior entities and objects proceeds concur-rently with his own. That which concerns the develop-ment of other entities will then follow the discussion of man's progress.

When the development of the Saturn consciousness was completed, there occurred one of the long rest periods (a *Pralaya*) mentioned above. After this there developed out of the heavenly body of man what in mystery science is called the "Sun." On the Sun, the human beings again emerged from their sleep. The previously developed Saturn consciousness was present in them as a predisposi-tion. First they again developed it from this germ. One can say that on the Sun man repeated the condition of Saturn before ascending to a higher one. However, it is not a simple repetition which is meant here, but one in another form. These transformations of forms will be dis-cussed later when we deal with the smaller cycles. At that time the differences between the individual "repeti-

tions" will also become apparent. Now we shall describe
only the development of consciousness.

After the repetition of the Saturn condition, the "Sun
consciousness" of man appears. This is somewhat clearer
than the preceding consciousness, but on the other hand
it has lost something with respect to broadness of vision.
In the deep, dreamless sleep of his present life, man has
a condition of consciousness similar to that which he
once had on the Sun. However, he who is not a clairvoyant
or a medium cannot perceive the objects and beings corresponding to the Sun consciousness. With the trance of
a medium reduced to this condition, and with the higher
consciousness of the true clairvoyant, the case here is
similar to what has been said with respect to the Saturn
consciousness.

The extent of the Sun consciousness is limited to the
Sun and the heavenly bodies most closely connected with
it. It is only these and their events which the inhabitant
of the Sun can experience as—to use once again the
simile employed above—man of today experiences his
heartbeat. In this way the inhabitant of Saturn could
also participate in the life of those heavenly bodies which
did not belong to the immediate sphere of Saturn.

When the Sun stage has passed through the appropriate subordinate cycles, it also enters a period of rest.
From this the heavenly body of man awakes to its "Moon
existence." Before ascending higher, again man passes
through the Saturn and Sun stage in two smaller cycles.
Then he enters his Moon consciousness. One can more

easily form an idea of the latter, for there is a certain
similarity between this stage of consciousness and a sleep
filled with dreams. It must be explicitly stated that here
again one can only speak of a *similarity*, not of an iden-
tity. It is true that the Moon consciousness is composed
of images such as appear in dreams, but these images
correspond to the objects and events around man in a way
similar to the ideas of the present "clear consciousness
of day." But everything in this correspondence is still dull,
in fact, imagelike. One can represent this state of affairs
to oneself in approximately the following way. Assume
that a Moon-being comes near an object, let us say near
salt. (Of course, at that time there was no "salt" in its
present form, but after all, in order to be understood,
one must remain in the area of images and similes.)
This Moon-being—the precursor of present-day man—does
not perceive an object with spatial extension and a definite coloring and form outside itself; instead, the approach
to this object causes a certain image—similar to a dream
image—to arise as it were within this being. This image
has a certain coloring which depends on the characteristics of the object. If the object is agreeable to the being
and useful for its existence, the coloring is light in yellow
nuances, or in green; if the object is disagreeable or is
one which is harmful to the being, a blood-like, reddish
color nuance appears. The clairvoyant also sees in this
way today, only he is fully conscious during this seeing, while the Moon inhabitant had only a dreamlike, dim
consciousness. The images appearing "within" these inhabitants had an exactly defined relationship to the en-

vironment. There was nothing arbitrary in them. It was possible to direct oneself by them; one acted under the impression of these images as today one acts under the impression of sensory perceptions.

The development of this dreamlike consciousness—the third principal stage—was the task of the "Moon cycle." When the "Moon" had passed through the appropriate "small cycles," a period of rest (*Pralaya*) again occurred. After this, the "Earth" emerged from the darkness.

T HE FOURTH principal stage of human development is lived on earth. This is that condition of consciousness in which man finds himself at present. But before he attained it, he, and with him the whole earth, first had to repeat successively the Saturn, Sun, and Moon stages in three smaller cycles (the so-called "rounds" of theosophical literature). Man now lives in the fourth earth cycle. He has already advanced a little past the middle of this cycle. At this stage of consciousness man no longer perceives in a dreamlike manner the images which arise in his soul through the influence of his environment only, but objects appear to him "outside in space." On the Moon and also during the stages of repetition on earth, there arose for example, a colored image in his soul when a particular object came near him. All of consciousness consisted of such images, tones, and so forth, which flowed and ebbed in the soul. Only with the appearance of the fourth condition of consciousness does color no longer appear merely in the soul, but on an external, spatially circumscribed object; sound is no longer merely an inner reverberating of the soul, but the resounding of an object in space. In mystery science therefore, one also calls this fourth, the earthly condition of consciousness, the "objective consciousness." It has been formed slowly and gradually in the course of development in the way that the physical organs of sense slowly arose and thus made perceptible the most diverse sensory qualities in external objects. Apart from the senses which are already developed, others exist in an as yet germinal state which will become fully developed in the subsequent earth period, and which will show the world of the senses in a diversity still greater than is the case today. The gradual growth of this earth consciousness has been described in the preceding pages, and in the discussions which are to follow this description will be amplified and supplemented in essential points.

The colored world, the sounding world, and so forth, which earlier man had perceived within himself, confronts him outside in space during his life on earth. But on the other hand, a new world appears within him: the world of ideas or thoughts. One cannot speak of ideas and thoughts in relation to the Moon consciousness. The latter consists solely of the images we have described. Around the middle of the development of earth—although this state of affairs was already preparing itself at a somewhat earlier time—there developed in man the capacity to form ideas and thoughts about objects. This capacity constitutes the basis for memory and for self-consciousness. Only conceptualizing man can develop a memory of that which he has perceived; and only thinking man reaches the point where he differentiates himself from his environment as an independent, self-conscious being, where he recognizes himself as an "I." The first three stages we have described were stages of consciousness; the fourth is not only consciousness, but *self-consciousness*.

But within the self-consciousness, the present-day life of thoughts, there is already developing a disposition toward still higher states of consciousness. Man will live through these states of consciousness on the next planets into which the earth will change after its present form. It is not absurd to say something about these future conditions of consciousness, and therewith about life on the following planets. For in the first place, the clairvoyant—for certain reasons which are to be given elsewhere—strides ahead of his fellows in his development. Thus those states of consciousness which all of mankind must attain with the advance of planetary development are already developing in him at this time. In the consciousness of the clairvoyant one finds an image of the future stages of mankind. Moreover, the three subsequent conditions of consciousness are now already present in all men in germinal states; and clairvoyant research has means for indicating what will emerge from these germinal states.

When it is said that the clairvoyant is already developing in himself the states of consciousness to which in future all of mankind will advance, this must be understood with one restriction. The clairvoyant, for example, is developing a seeing in the spiritual world today which in future will appear in man in a physical way. But this

future physical condition of man will be a faithful likeness of the corresponding contemporary spiritual one in
the clairvoyant. The earth itself is going to develop, and
therefore quite different forms from those which exist
today will appear in its future physical inhabitants; but
these physical forms are being prepared in the spiritual
and mental ones of today. For example, what the clairvoyant today sees in the form of a cloud of light and
color around the human physical body as a so-called
"aura," will later change into a physical form; and other
organs of sense than those of today will give the man
of the future the capacity to perceive other forms. However, already today the clairvoyant sees the spiritual mod-
els of the later material entities with his spiritual senses
(thus for example, the aura). *A view into the future* is
possible for him, although it is very difficult to give an
idea of the character of this view through the language
of today and for present-day human conceptions.

The conceptions of the present state of consciousness
are shadowy and pale in comparison with the colorful
and sounding objects of the external world. Man therefore speaks of conceptions as of something which is "not
real." A "mere thought" is contrasted with an object or
a being which is "real" because it can be perceived through
the senses. But conceptions and thoughts bear within
themselves the potentiality of again becoming real and
imagelike. If man speaks of the conception "red" today
without having a red object before him, then this conception is, as it were, only a shadow image of real "red-
ness." Later, man will reach the point where he can not

only let the shadowy conception of the "red" arise in his soul, but where, when he thinks "red," "red" will actually be before him. He will be able to *create* images, not merely conceptions. Thereby something will be achieved by him similar to that which already existed for the Moon consciousness. But the images will not ebb and flow in him like dreams; instead he will evoke them in full *self-consciousness,* as he does today's conceptions. The thought of a color will be the color itself; the conception of a sound will be the sound itself, and so forth. In the future, a world of images will flow and ebb in the soul of man through his own power, whereas during the Moon existence such a world of images filled him without his acting. In the meantime the spatial character of the objective external world will not disappear. The color which arises together with the conception of color will not be merely an image in the soul but will appear in outside space. The consequence of this will be that man will be able to perceive beings and objects of a higher kind than those of his present environment. These are objects and beings which are of a more delicate spiritual and soul nature, hence they do not clothe themselves in the objective colors which are perceptible to the present physical sense organs; however, these are objects and beings which will reveal themselves through the more delicate spiritual and mental colors and sounds which the man of the future will be able to create from his soul.

Man is approaching a condition in which he will have a *self-conscious image consciousness** appropriate for such

---

* The combination, "self-conscious image consciousness," may appear

perceptions. On the one hand, the coming development of earth will raise the present life of conceptions and thoughts to an ever higher, more delicate, and more perfect condition; on the other hand, the self-conscious image consciousness will gradually develop itself during this time. The latter, however, will attain full life in man only on the next planet into which the earth will transform itself, and which is called "Jupiter" in mystery science. Then man will be able to enter into intercourse with beings which are completely hidden from his present sensory perception. It will be understood that not only does the life of perception thereby become totally different, but that actions, feelings, and all relations to the environment, are completely transformed. While today man can consciously influence only sensory beings, he will then be able to act consciously on very different forces and powers; he himself will receive what to him will be fully recognizable influences from very different realms than at present. At that stage there can no longer be any question of birth and death in the present sense. For "death" occurs only because the consciousness has to depend on an external world with which it enters into communication through the physical sense organs. When these physical sense organs fail, every relation to the environment ceases. That is to say, the man "has died." However, when his soul is so far advanced that it does not receive the influences of the outside world through physical instruments, but receives them through the im-

---

odd, but it probably best expresses the state of affairs. If one wished, one could also say, "image self-consciousness."

ages which the soul creates out of itself, then it will have
reached the point where it can regulate its intercourse
with the environment independently, that is, its life will
not be interrupted against its will. It has become lord
over birth and death. All this will come to be with the
developed self-conscious image consciousness on "Jupiter."
This state of the soul is also called the "psychic consciousness."

The next condition of consciousness to which man
develops on a further planet, "Venus," is distinguished
from the previous one by the fact that the soul can now
create not only images, but also objects and beings. This
occurs in the *self-conscious object consciousness* or supra-
psychic consciousness. Through the image consciousness
man can perceive something of supersensible beings and
objects, and he can influence them through the awakening
of his image conceptions. But in order for that to take
place which he desires of such a supersensible being, at
his instigation, this being must use its own forces. Thus
man is the ruler over images, and he can produce effects
through these images. But he is not yet lord over the
forces themselves. When his self-conscious object consciousness is developed, he will also be ruler over the
creative forces of other worlds. He will not only perceive
and influence beings, but he himself will create.

This is the course of the development of consciousness:
at first it begins dimly; one perceives nothing of other
objects and beings, but only the inner experiences (images) of one's own soul; then perception is developed.
At last the perceptive consciousness is transformed into

a creative one. Before the condition of earth goes over
into the life of Jupiter—after the fourth earthly cycle—
there are three more smaller cycles to be passed through.
These serve for the further perfection of the consciousness of earth in a manner to be described in the following
essays, when the development of the smaller cycles and of
their subdivisions will be described for all seven planets.
When, after a period of rest (*Pralaya*) , earth has changed
into Jupiter, and when man has arrived on the latter
planet, then the four preceding conditions—Saturn, Sun,
Moon, and earth condition—must again be repeated during four smaller cycles; and only during the fifth cycle
of Jupiter does man attain the stage which has been
described above as the real Jupiter consciousness. In a
corresponding manner does the "Venus consciousness"
appear during the sixth cycle of Venus.

A fact which will play a certain role in the following
essays will be briefly indicated here. This concerns the
speed with which the development on the different planets
takes place. For this is not the same on all the planets.
Life proceeds with the greatest speed on Saturn, the
rapidity then decreases on the Sun, becomes still less on
the Moon and reaches its slowest phase on the earth. On
the latter it becomes slower and slower, to the point at
which self-consciousness develops. Then the speed increases again. Therefore, today man has already passed
the time of the greatest slowness of his development. Life
has begun to accelerate again. On Jupiter the speed of
the Moon, on Venus that of the Sun will again be attained.

The last planet which can still be counted among the

series of earthly transformations, and hence follows Venus, is called "Vulcan" by mystery science. On this planet the provisional goal of the development of mankind is attained. The condition of consciousness into which man enters there is called "piety" or spiritual consciousness. Man will attain it in the seventh cycle of Vulcan after a repetition of the six preceding stages. Not much can be publicly communicated about the life on this planet. In mystery science one speaks of it in such a way that it is said, "No soul which, with its thinking is still tied to a physical body, should reflect about Vulcan and its life." That is, only the mystery students of the higher order, who may leave their physical body and can acquire super-sensible knowledge outside of it, can learn something about Vulcan.

The seven stages of consciousness are thus expressed in the course of the development of mankind in seven planetary developments. At each stage, the consciousness must now pass through seven subordinate conditions. These are realized in the smaller cycles already mentioned. (In theosophical writings these seven cycles are called "rounds.") These subordinate states are called "conditions of life" by the mystery science of the Occident, in contrast with the superordinated "conditions of consciousness." Or, one says that each condition of consciousness moves through seven "realms." According to this calculation, one must distinguish seven times seven in the whole development of mankind, that is, forty-nine small cycles or "realms" (according to common theosophical usage, "rounds"). And again, each small cycle has to pass through seven yet smaller ones, which are called "conditions of form" (in theosophical language, "globes"). For the full cycle of humanity this amounts to seven times forty-nine or three hundred and forty-three different "conditions of form." ← ⟶ 343

The following discussions which deal with this development, will show that a survey of the whole is not as complicated as might at first appear at the mention of the number three hundred and forty-three. It will become apparent how man can only truly understand himself when he knows his own development.

IN ONE of the preceding descriptions, the great development of humanity through the seven stages of consciousness from Saturn to Vulcan has been compared with the progress through life between birth and death, through infancy, childhood, and so on, to old age. One can extend this comparison further. As among contemporary humanity, men of different ages do not only follow upon one another, but also exist side by side, so it is with the development of the stages of consciousness. The aged man, the mature man or the mature woman, the youth, travel through life *side by side*. Thus the ancestors of man existed on Saturn not only as beings with the dull Saturn consciousness, but also along with these as beings which had already developed the higher stages of consciousness. When the Saturn development began, there already existed natures with Sun consciousness, others with image consciousness (Moon consciousness), those with a consciousness similar to the present consciousness of man, then a fourth kind with self-conscious (psychic) image consciousness, a fifth with self-conscious (supra-psychic) object consciousness, and a sixth with creative (spiritual) consciousness. This does not exhaust the series of beings. After the Vulcan stage, man will develop yet further, and will ascend to still higher levels of consciousness. As the external eye looks into misty gray distances, so the inner eye of the seer looks upon five more forms of consciousness, as far off as distant spirits, of which a description, however, is quite impossible. In all, one can speak of *twelve* stages of consciousness.

The Saturn man was surrounded by eleven other kinds of beings. The four highest had had their tasks on levels of development which preceded the life of Saturn. When this life began they had already arrived at such a high stage of development that their further existence took place in worlds which lie beyond the realms of man. Therefore, we cannot and need not speak of them here.

The other kinds of beings, however—seven of them in addition to the Saturn man—are all concerned in the human development. In this they act as creative powers, performing their services in a way which will be described in the following pages.

When the Saturn development began, the most sublime of these beings already had attained a level of consciousness which man will reach only after his Vulcan life, that is, a high creative (supra-spiritual) consciousness. These "creators," too, once had to pass through the stages of man. This took place on heavenly bodies which preceded Saturn. However, the connection of these beings with the development of mankind lasted until the middle of the life of Saturn. Because of their sublime, delicate body of rays, in mystery science they are called "Radiating Lives" or "Radiating Flames." Because the substance of which this body consisted had a remote resemblance to the will of man, they are also called "Spirits of Will."

These spirits are the creators of the man of Saturn. From their bodies they pour the substance which becomes the carrier of the human Saturn consciousness. The period of development during which this takes place is called the first small Saturn cycle. (In the language of theosophy, this is the "first round.") The material body which man receives in this way is the first rudiment of his later physical body. One can say that the germ of the physical human body is planted during the first Saturn cycle by the Spirits of Will, and that at that time this germ has the dull Saturn consciousness.

This first smaller Saturn cycle is followed by six others. In the course of these cycles man does not attain a higher degree of consciousness. But the material body which he has received is further elaborated. The other kinds of beings indicated above participate in this elaboration in the most diverse ways.

After the "Spirits of Will" there follow beings with a creative (spiritual) consciousness, similar to that which man will attain on Vulcan. They are called "Spirits of Wisdom." Christian mystery science calls them "Dominions" (*Kyriotetes*), while it calls the "Spirits of Will," "Thrones."* During the second cycle of Saturn they ad-

---

* He who really knows Christian doctrine is aware that the conceptions of these spiritual beings superordinated to man form an integral

vance their own development to some extent, and at the same time work on the human body in such a way that a "wise arrangement," a rational structure is implanted in it. To be more exact, their work on man already begins shortly after the middle of the first cycle and is completed in about the middle of the second.

The third kind of spirits with the self-conscious (suprapsychic) object consciousness is called "Spirits of Motion" or of "Activity." In Christian mystery science they are called "Principalities" (*Dynamis*). (In theosophical literature, the expression *Mahat* is to be found for them.) From the middle of the second Saturn cycle onward they combine with the progress of their own development, the further elaboration of the human material body, in which they implant the capacity of movement and of forceful activity. This task comes to a conclusion around the middle of the third Saturn cycle.

After this point, the work of the fourth kind of beings, the so-called "Spirits of Form," begins. They have a self-

---

part of it. Only for some time they have been lost by an externalized religious teaching. The one who really enters into these matters and looks deeper will realize that there is not the slightest reason for Christianity to combat mystery science, but that on the contrary the latter is in complete harmony with true Christianity. If, for the sake of their Christianity, the theologians and teachers of religion were to agree to study mystery science, they would have to recognize in it their best helper and means of advancement today. But many theologians think in a completely materialistic manner, and it is characteristic that in a popular publication intended for the furthering of a knowledge of Christianity, today one can even read that "Angels" are for "children and nurses." Such a statement springs from a complete misunderstanding of the true Christian spirit. Only the man who sacrifices true Christianity to a supposedly advanced "science" can make such a statement. But the time will come when a higher science will go beyond the childishness of such utterances to matters of real importance.

conscious image consciousness (psychic consciousness).
Christian esoteric teaching names them "Powers" (*Exusiai*).Through their work, the human material body,
which previously was a kind of mobile cloud, receives a
bounded, plastic form. This activity of the "Spirits of
Form" is completed around the middle of the fourth Saturn cycle.

Then follows the activity of the "Spirits of Darkness,"
which are also called "Spirits of Personality" or of "Selfhood" (Egoism). At this stage they have a consciousness
similar to the present human earthly consciousness. They
inhabit the formed human material body as "souls" in a
way similar to that in which the human soul inhabits its
body today. They implant a kind of sensory organs in the
body, which are the germs of the sensory organs which
later develop in the human body in the course of the
development of earth.

One must realize, however, that these "sensory germs"
are still substantially different from the present sensory
instruments of man. Earth man could not perceive
through such "sensory germs." For him, the images of the
sensory instruments must first pass through a more refined
ether body, which forms on the Sun, and through an astral
body, which owes its existence to the Moon development.
(All this will become clear in the following chapters.)
But the "Spirits of Personality" can treat the images of
the "sensory germs" through their own soul in such a way
that, with their aid, they can perceive external objects,
as does man during his earthly development. In their
work on the human body, the "Spirits of Personality" pass

through their own "stage of humanity." Thus they are *men* from the middle of the fourth to the middle of the fifth Saturn cycle.

These spirits implant selfhood, egoism in the body of man. Since they only attain their stage of humanity on Saturn, they remain connected with the development of mankind for a long time. Thus they have important work to perform on man in subsequent cycles as well. This work always acts as an inoculation with selfhood. The degenerations of selfhood into selfishness must be ascribed to their activity, while on the other hand they are the originators of all of man's independence. Without them man would never have become a self-enclosed entity, a "personality." Christian esoteric teaching uses the expression "Primal Beginnings" (*Archai*) for them, and in theosophical literature they are designated as *Asuras*.

The work of these spirits is succeeded around the middle of the fifth Saturn cycle by that of the "Sons of Fire," who, at this stage, still have a dull image consciousness, similar to the Moon consciousness of man. They attain the stage of humanity only on the next planet, the Sun. Their work here is therefore to a certain degree still unconscious and dreamlike. But it is through them that the activity of the "sensory germs" from the previous cycle is enlivened. The light images produced by the "fire spirits" shine outward through these sensory germs. The ancestor of man is thereby elevated to a kind of shining entity. While the life of Saturn is otherwise dark, man now shines in the general darkness.

The "Spirits of Personality" on the other hand, were

still awakened to their human existence in this general darkness.

The human being himself can make no use of his luminosity on Saturn. The luminosity of his sensory germs could not express anything in itself, but through it other more exalted beings are given the possibility to reveal themselves to the life of Saturn. Through the sources of light of the ancestors of man, these beings radiate something of their nature down to the planet. These are exalted beings from among those four ranks of which it has been said above that they have grown beyond all connection with human existence in their development. Without any necessity for them to do it, they now radiate something of their nature out of "free will." Christian esoteric teaching here speaks of the revelation of the *Seraphime* (Seraphim) , the "Spirits of Love." This condition lasts until the middle of the sixth Saturn cycle.

After this begins the work of those beings which at this stage have a dull consciousness such as is found in man today when he is in a deep, dreamless sleep. These beings are the "Sons of Twilight," the "Spirits of Dusk." (In theosophical writings they are called *Lunar Pitris* or *Barhishad-Pitris*.) They attain the stage of humanity only on the Moon. On earth they, as well as their predecessors, the Sons of Fire, have already grown beyond the stage of humanity. On earth they are higher beings which Christian esoteric teaching calls "Angels" (*Angeloi*) , while for the Sons of Fire it uses the expression "Archangels" (*Archangeloi*) . These Sons of Twilight develop in the ancestor of man a kind of understanding, of which however, in his dull consciousness, he himself cannot yet make use. Through this understanding, exalted entities now again reveal themselves, as previously the Seraphim did through the sensory germs. Through the human bodies, understanding is now poured out over the planet by those spirits whom Christian esoteric teaching calls *Cherubime* (Cherubim).

Around the middle of the seventh Saturn cycle a new activity begins. Man has now reached the point where he can work unconsciously on his own material body. Through his activity in the utter dullness of Saturn existence, man produces the first germinal predisposition to the true "spirit man," who reaches his full development only at the end of the development of mankind. In theosophical literature this is called *Atma*. It is the highest member of the so-called monad of man. In itself it would be quite dull and unconscious at this stage. But as the Seraphim and the Cherubim reveal themselves out of their free will in the two preceding human stages, so the Thrones now reveal themselves, those beings who, at the very beginning of Saturn existence, radiated the human body out of their own nature. The germinal predisposition of "spirit man" (*Atma*) is completely penetrated by the power of these Spirits of Will and retains this power through all subsequent stages of development. In his dull consciousness at this stage man as yet cannot realize anything of this germinal predisposition; but he develops further, and later this germinal predisposition becomes clear to his own consciousness.

This work is not yet completed at the end of the life

of Saturn; it continues into the first Sun cycle. One should consider that the labor of the higher spirits which has been described here does not coincide with the beginning and end of a smaller cycle (of a round), but that it continues from the middle of one to the middle of the next. Its greatest activity is developed *in the periods of rest between the cycles*. It increases from the middle of a cycle (*Manvantara*) onward, becomes strongest in the middle of a period of rest (*Pralaya*), and then ebbs in the next. (It has already been mentioned in the preceding chapters that life by no means ceases during the periods of rest.)

From the above it also becomes apparent in what sense Christian esoteric science says that in the "beginning of time" the Seraphim, Cherubim, and Thrones first revealed themselves.

With this, the course of Saturn has been followed to the time where its life develops through a period of rest into that of the Sun. Of this we shall speak in the following discussions.

*

For the sake of greater clarity, here we shall give a summary of the facts of development of the first planet.

I. This planet is the one on which the dullest human consciousness develops (a deep trance consciousness). Together with this, the first rudiment of the physical human body develops.

II. This development passes through seven subsidiary stages (smaller cycles or "rounds"). At each of these stages higher spirits begin their work on the development of the human body, namely in the

1st cycle, the Spirits of Will (Thrones),

2nd cycle, the Spirits of Wisdom (Dominions),

3rd cycle, the Spirits of Motion (Principalities),

4th cycle, the Spirits of Form (Powers),

5th cycle, the Spirits of Personality (Primal Beginnings),

6th cycle, the Spirits of the Sons of Fire (Archangels),

7th cycle, the Spirits of the Sons of Twilight (Angels).

III. In the fourth cycle, the Spirits of Personality raise themselves to the stage of humanity.

IV. From the fifth cycle onward, the Seraphim reveal themselves.

V. From the sixth cycle onward, the Cherubim reveal themselves.

VI. From the seventh cycle onward, the Thrones, the true "creators of man," reveal themselves.

VII. Through the latter revelation, there develops in the seventh cycle of the first planet, the predisposition to the "spiritual man," to *Atma*.

AFTER THE GREAT cosmic era of Saturn, which has been described in the preceding pages, there follows that of the Sun. Between them lies a period of rest (*Pralaya*). During this period, everything human which has developed on Saturn takes on a character which stands in the same relation to the subsequently to be developed Sun man as the seed to the plant which emerges from it. Saturn man, as it were, has left behind his seed, which is sunk in a kind of sleep, after which it will develop into Sun man.

Man now passes through his second stage of consciousness on the Sun. It resembles that into which today man sinks during a calm and dreamless sleep. This condition, which interrupts man's state of wakefulness today, is a remainder, as it were, a memory of the time of the Sun development. One can also compare it with that dull state of consciousness in which the world of plants exists today. As a matter of fact, in the plant one must see a sleeping being.

In order to understand the development of mankind, one must realize that in this second great cycle the Sun was

still a planet, and that only later did it advance to the
existence of a fixed star. In the sense of mystery science,
a fixed star is one which sends life forces to one or several
planets situated at a distance from it. During the second
cycle this was not yet the case with the Sun. At that time
it was still united with the beings to which it gave force.
These beings—and also man at his level of development
of that time—still lived on it. A planetary earth, separated
from Sun and Moon, did not exist. Everything in the way
of substances, forces, and beings which exists on and in
the earth today, and everything which now belongs to
the Moon, was still within the Sun. It formed a part of *its*
substances, forces, and beings. Only during the next
(third) great cycle did that detach itself from the Sun
which in mystery science is called the *Moon*. This is not
the present moon, but the predecessor of our earth, its
previous embodiment (reincarnation), as it were. This
Moon became the earth, after it in turn had detached from
its substance and cast off what one today designates as
moon. In the third cycle two bodies thus existed in place
of the former planetary Sun, namely, the fixed star Sun
and the split-off planetary Moon. Man and the other beings
which had developed as man's companions during the
course of the Sun, had been taken out of the Sun along
with the Moon. The Sun now provided the Moon beings
from the outside with those forces which they had previously obtained directly from it as their dwelling-place.

After the third (Moon) cycle there occurred another
period of rest (*Pralaya*). During this period the two separate bodies (Sun and Moon) became united and to-

gether passed through the condition of the sleeping seed. In the fourth cyclic period, Sun and planetary Moon at first emerged from the obscurity of sleep as *one* body. During the first half of this cycle our earth, along with man and his companions, split off from the Sun. A little later it cast off the present moon, so there now exist three members as descendants of the former Sun planet.

On the Sun planet, man and the other beings mentioned in the course of the discussion of Saturn passed through another stage of their development in the second great cosmic era. The rudiment of the later physical body of man, which had gradually developed on Saturn, emerges like a plant from the seed at the beginning of the Sun cycle. But here it does not remain in the same state in which it was previously. It is permeated by a second, more delicate, but in itself more powerful body, the ether body. While the Saturn body of man was a kind of automaton (quite lifeless), now, through the ether body which gradually permeates it completely, it becomes an animated being. Man thereby becomes a kind of plant. His appearance, however, is not that of the plants of today. Rather in his forms he already somewhat resembles present-day man. But, the rudiment of the head like the plant root of today, is turned downward, toward the center of the Sun, and the rudiments of the feet are turned upward like the blossom of the plant. This plant-man organism has as yet no capacity of voluntary movement.*

* For someone who clings to the sensory perceptions of today, it will of course be difficult to imagine that man lived as a plant being in the Sun itself. It seems inconceivable that a living being could exist in the physical conditions which must be assumed for this state. But it is only

But man only develops into this form during the second of the seven smaller cycles (rounds) through which the Sun passes. For the duration of the first of these small cycles there is as yet no ether body in the human organism. Everything which occurred during the Saturn era is then repeated in brief. The physical body of man still retains its automatic character, but it changes its previous form somewhat. If it were to remain as it was on Saturn, it would not be capable of harboring an ether body. It is changed in such a way that it can become a carrier of this body. During the following six cycles the ether body is developed further and further, and through its forces, which act on the physical body, the latter also gradually receives a more and more perfect form.

The work of transformation which is performed on man here is carried out by the spirits which have already been mentioned in connection with man in our discussion of the Saturn development.

Those spirits which are called "Radiating Lives" or "Flames" (in Christian esoteric science, "Thrones"), are now no longer in question. They have performed their labor in this respect during the first half of the first Saturn cycle. What can be observed during the first Sun cycle (round) is the labor of the "Spirits of Wisdom" (Dominions or *Kyriotetes* in Christian esoteric doctrine). They have intervened in the development of man around the middle of the first Saturn cycle (see the previous chap-

---

a plant of *today* which is adapted to the present physical earth. It has only developed in this way because its environment is a corresponding one. The Sun plant-being existed under other conditions of life, which corresponded to the physical solar conditions of that time.

ter). They now continue their labor during the first half of the first Sun cycle by repeating in successive stages the wise arrangement of the physical body. A little later this labor is joined by that of the "Spirits of Motion" (*Dynamis* in Christianity, *Mahat* in theosophical literature). Thereby that period of the Saturn cycle is repeated during which the human body received the capacity of motion. It thus again becomes mobile. In the same way the "Spirits of Form" (*Exusiai*), those of "Darkness" (in Christianity, *Archai,* in theosophy, *Asuras*), then the "Sons of Fire" (Archangels), and finally the "Spirits of Twilight" (Angels, *Lunar Pitris*) successively repeat their labors. Therewith we have characterized six smaller periods of the first course of the Sun (of the first solstice).

In a seventh of these smaller periods the "Spirits of Wisdom" again intervene. While in their preceding period of labor they had given a wise structure to the human body, they now bestow on the limbs, which have become mobile, the capacity to render their motion a wisdom-directed one. Previously it was only the structure which was an expression of inner wisdom; now the motion too becomes such an expression. With this, the first Sun cycle attains its end. It consists of seven successive smaller cycles, of which each one is a short repetition of a Saturn cycle (a Saturn round). In theosophical literature one has become accustomed to calling these seven smaller cycles, which make up a so-called "round," "globes." (A round thus takes place in seven "globes.")

Now, after a period of rest (*Pralaya*), the first Sun cycle is succeeded by the second. The individual "smallest cycles" or "globes" will be discussed in detail later; at present we shall proceed to the subsequent course of the Sun cycle.

At the end of the first, the human body is already prepared for the reception of the ether body, because the "Spirits of Wisdom" have given him the possibility of wisdom-filled motion.

In the meantime however, these "Spirits of Wisdom" themselves have developed further. Through the labor which they have performed, they have become capable of pouring their substance out of themselves just as the "Flames" poured theirs out in the beginning of the Saturn cycle, thereby giving the physical body its material basis. The substance of the "Spirits of Wisdom" is the "ether," that is, mobile and power-filled wisdom, in other words, "life." The ether or life body of man is thus an emanation of the "Wisdom Spirits."

This emanation continues until around the middle of the second Sun cycle, when the "Spirits of Motion" can again begin with a new activity. Their labor previously could only extend to the physical body of man; now it is transferred to the ether body and implants a powerful activity in it. This continues until the middle of the third Sun cycle. Then the action of the "Spirits of Form" begins. Through them the ether body, which before had had only a cloudlike mobility, receives a definite shape (form).

In the middle of the fourth course of the Sun, these "Spirits of Form" receive a consciousness like that which man will have on "Venus," the second planet on which

he will appear after his earthly existence. This is a supra-
psychic consciousness. These spirits attain this as a fruit
of their activity during the third and fourth course of
the Sun. Thereby they acquire the capacity to transform
the sensory germs developed during and after the Saturn period, and which until this time were only physical
instruments, into *animated senses* by means of the ether.

Through a similar process the "Spirits of Darkness"
(in Christianity, *Archai,* in theosophy, *Asuras*) have at
this time attained the level of psychic consciousness,
which man will develop only on Jupiter as conscious
image consciousness. Thereby they become capable of
acting consciously from the astral world. Now the ether
body of a being can be influenced from the astral world.
The "Spirits of Darkness" did this with respect to the
ether body of man. They now implanted in it the spirit
of selfhood (independence and selfishness), as they had
previously done in the physical body. One can see how
these spirits implanted egoism in all the members of
the human entity in turn.

At the same time the "Sons of Fire" attained the stage
of consciousness which man today possesses as his waking
consciousness. One can say of them that they now become *men.* Now they can make use of the physical human
body for a kind of intercourse with the outside world.
In similar fashion the "Spirits of Personality" made use
of the physical body from the middle of the fourth Saturn cycle on. But they had used the sensory germs for
a kind of perception. The nature of the "Sons of Fire,"

however, is such that they pour the warmth of their soul out into their environment. The physical human body is now so far advanced that they can do this through it. Their warmth acts approximately like the warmth of the hen on the egg which she is hatching, that is, it has a life-awakening power. Everything of such a life-awakening power that lies in man and in his companions was implanted into the ether body at that time by the Sons of Fire. We are dealing here with the origin of that warmth which is a condition for the reproduction of all living beings. Later it will become apparent what kind of a transformation this power of warmth went through when the Moon split off from the Sun.

Around the middle of the fifth cycle the "Sons of Fire" have developed so far that they can inoculate the ether body with the capacity which they previously exercised through the physical human body. They now relieve the "Spirits of Personality" in the work on this ether body, which thereby becomes the initiator of a reproductive activity.

In this period they abandon the physical body to the Sons of Twilight (in Christianity, Angels, in theosophy, *Lunar Pitris*) . In the meantime, the latter have acquired a dull image consciousness such as man will have on the Moon. On Saturn they had given the ancestor of man a kind of organ of understanding. Now they further develop the physical instruments of the human spirit, which he will consciously use at later stages of his development. Thereby, through the human body the Seraphim can already reveal themselves on the Sun before the middle of the fifth cycle in a more complete manner than was possible on Saturn.

From the middle of the sixth course of the Sun onward, man himself is so far advanced that he can unconsciously work on his physical body. In this respect he now relieves the "Sons of Twilight." Through this activity in dullness, he creates the first germinal predisposition to the living spiritual being, which one calls life-spirit (*Buddhi*). Only at later stages of his development will he become conscious of this spirit of life. As from the seventh Saturn cycle onward, the Thrones voluntarily poured their power into the predisposition to spirit-man which was formed at that time, so the Cherubim now pour out their wisdom, which thenceforward is preserved for the life-spirit of man through all subsequent stages of development. From the middle of the seventh course of the Sun onward, the germ of spirit-man (*Atma*), already formed on Saturn, appears again. It combines with the life-spirit (*Buddhi*), and the animated monad (*Atma-Buddhi*) thus comes into being.

While man works unconsciously on his physical body in this time, the Sons of Twilight take over what must now be done on the ether body in order to develop it further. In this respect they are the successors of the Sons of Fire. They radiate the images of their consciousness into this ether body and thereby, in a kind of dreamlike condition, enjoy the power of reproduction of this body, which has been stimulated by the Sons of Fire. By this, they prepare the development of the pleasure in this power, which later (on the Moon) appears in man and in his fellow-beings.

On Saturn, man's physical body had been formed. The latter was completely lifeless at that time. Such a lifeless body is called mineral by mystery science. One can therefore also say that on Saturn man was mineral, or he passed through the mineral realm. This human mineral did not have the form of a present-day mineral. Minerals as they are at present, did not yet exist at that time.

As has been shown, this human mineral which re-emerged from the obscurity of sleep as from a germ, was animated on the Sun. It became a human plant; man passed through the realm of plants.

But not all human minerals are animated in this manner. This could not have happened, for the plant man needed the mineral basis for his life. As today there can be no plants without a mineral realm from which they take in their substances, so was it on the Sun with respect to the plant man. For the sake of his further development, the latter had to leave a portion of the human rudiments behind at the level of the minerals. Since on the Sun conditions were quite different from those of Saturn, these minerals which had been thrust back assumed forms quite different from those they had had on Saturn. Thus, alongside the human plant realm, a second province, a special mineral realm came into being. It can be seen that man ascends into a higher realm by thrusting a part of his companions down into a lower.

We shall see this process repeating itself many times in the subsequent stages of development. It corresponds to a fundamental law of development.

*

Here again, for the sake of greater clarity, we shall give a summary of the facts of development on the Sun.

I. The Sun is the planet on which develops the second human condition of consciousness, that of dreamless sleep. The physical human body rises to a kind of plant existence through the incorporation of an ether body into it.

II. This development passes through seven subsidiary stages (smaller cycles or "rounds").

1. In the first of these cycles the stages of development of Saturn are repeated, with respect to the physical body, in a somewhat altered form.

2. At the end of the *first cycle* begins the pouring out of the ether body by the "Spirits of Wisdom."

3. In the middle of the *second cycle,* the work of the "Spirits of Motion" on this body begins.

4. In the middle of the *third cycle* the action of the "Spirits of Form" on the ether body has its beginning.

5. From the middle of the *fourth cycle* onward, this body receives selfhood through the "Spirits of Personality."

6. In the meantime, the physical body has advanced so far through the action of the forces which have been working on it since earlier periods that from the *fourth*

cycle onward the "Spirits of Fire" can elevate themselves to humanity through it.

7. In the middle of the *fifth cycle* the "Spirits of Fire," which have previously passed through the stage of humanity, take over the work on the ether body. The "Sons of Twilight" are active in the physical body at this time.

8. Around the middle of the *sixth cycle* the work on the ether body is transferred to the "Sons of Twilight." Man himself now works on the physical body.

9. In the course of the *seventh cycle* the animated monad has come into existence.

In THE UNIVERSAL ERA of the *Moon,* which follows that of the Sun, man develops the third of his seven states of consciousness. The first had developed during the seven Saturn cycles, the second during the Sun development; the fourth is that which man is at present developing during the course of the earth; three others will come into being on subsequent planets. The condition of consciousness of Saturn man cannot be compared with any state of consciousness of present-day man, for it was duller than that of dreamless sleep. The Sun consciousness, however, can be compared to this condition of dreamless sleep, or to the present consciousness of the sleeping plant world. But in all these instances one is dealing only with similarities. It would be quite erroneous to think that in the great universal eras anything repeats itself in a completely identical manner.

It is to be understood in this way if the Moon consciousness is now compared with one with which it has some similarity, namely with that of dream-filled sleep. Man attains the so-called image consciousness on the

Moon. The similarity consists in that in the Moon consciousness as well as in dream consciousness, *images* arise within a being which have a certain relation to objects and beings of the outside world. But these images are not likenesses of these objects and beings as in present-day man when he is awake. The dream images are echoes of the experiences of the day, or symbolical expressions for events in the dreamer's environment, or for what is taking place in the interior of the dreaming person. Examples of these three types of dream experiences are easy to give. First, everyone knows those dreams which are nothing but confused images of more or less remote daily experiences. An example of the second type would be if the dreamer thinks he perceives a passing train and then, upon awakening, realizes that it was the ticking of the watch lying beside him which was perceptible in this dream image. An example of the third kind is that it seems to someone that he is in a room where ugly animals are sitting on the ceiling, and upon awaking from this dream he realizes that it was his own headache which expressed itself in this way.

If one now wants to attain a conception of the Moon consciousness on the basis of such confused dream images, one must realize that while the image-like character is also present there, complete regularity instead of confusion and arbitrariness prevails. It is true that the images of the Moon consciousness have even less similarity than the dream images to the objects to which they are related, but on the other hand there is a complete *correspondence* of image and object. At present in the earth development,

the conception is a likeness of its object; thus for instance
the conception "table" is a likeness of the table itself.
This is not the case with the Moon consciousness. Assume,
for instance, that the Moon man approaches an object
which to him is pleasing or advantageous. Then a colored
image of a light tone arises in his soul; when something
harmful or displeasing comes near him, he beholds an
ugly, dark image. The conception is not a likeness, but
a *symbol* of the object which corresponds to it in a quite
definite and regular way. Hence the being which has such
symbolical conceptions can direct its life in accordance
with them.

The inner life of man's ancestor on the Moon thus
took its course in images which have the character of
the volatile, the floating, and the symbolical in common
with dreams of today, but are distinguished from these
dreams by their completely regular character.

The basis for the development of this image consciousness in man's ancestors on the Moon was the formation
of a third member in addition to the physical body and
the ether body. This third member is called the astral
body.

This formation, however, only occurred in the third
smaller Moon cycle—the so-called third Moon round.
The first two revolutions of the Moon must be seen
merely as a repetition of what took place on Saturn and
on the Sun. But this repetition must not be imagined
as a re-enactment of all the events which took place on
Saturn and on the Sun. That which repeats itself, namely
the development of a physical body and of an ether body,

at the same time is subject to such a transformation that in the third Moon cycle these two members of the nature of man can be united with the astral body, a union which could not have taken place on the Sun.

In the third Moon period—actually the process already starts around the middle of the second—the Spirits of Motion pour the astral element out of their own nature into the human body. During the fourth cycle—from the middle of the third onward—the Spirits of Form shape this astral body in such a way that its form, its whole organization can develop inner processes. These processes have the character of what at present in animals and man is called instinct, desire—or the appetitive nature. From the middle of the fourth Moon cycle onward, the Spirits of Personality begin with their principal task in the fifth Moon era: they inoculate the astral body with selfhood, as they have done in the preceding cosmic eras with respect to the physical and the ether body. But in order for the physical and the ether body to be so far advanced that they can harbor an independent astral body, at the time indicated, that is, in the middle of the fourth Moon cycle, they must first be brought to this point by the shaping spirits in the successive stages of development. This takes place in the following manner. The physical body is brought to the necessary maturity in the first course of the Moon (round) by the Spirits of Motion, in the second by those of Form, in the third by those of Personality, in the fourth by the Spirits of Fire, and in the fifth by those of Twilight. To be *exact,* this labor of the Spirits of Twilight takes place from the middle of the fourth Moon cycle onward, so that at the same time that the Spirits of Personality are engaged on the astral body the same is the case with the Spirits of Twilight with respect to the physical body.

In regard to the ether body the following is the case. Its necessary qualities are implanted in it in the first course of the Moon by the Spirits of Wisdom, in the second by those of Motion, in the third by those of Form, in the fourth by those of Personality, and in the fifth by those of Fire. To be exact, *this* activity of the Fire Spirits takes place concurrently with the labor of the Spirits of Personality on the astral body, that is, from the middle of the fourth course of the Moon, onward into the fifth.

If one considers the entire ancestor of man as he developed on the Moon at that time, there is this to be said: Starting from the middle of the fourth Moon cycle, man consists of a physical body in which the Sons of Twilight perform their labor, of an ether body in which the Spirits of Fire perform theirs, and finally of an astral body in which the Spirits of Personality perform theirs.

That the Spirits of Twilight work on the physical body of man in this period of development, means that they now rise to the level of *humanity,* as did the Spirits of Personality in the same cycle on Saturn and the Fire Spirits on the Sun. One must imagine that the "sensory germs" of the physical body, which by that time have become further developed, can be used by the Spirits of Twilight from the middle of the fourth course of the Moon onward in order to perceive external objects and events on the Moon. Only on the earth will man be so far advanced that, from the middle of the fourth cycle onward, he can make use of these senses. On the other hand, around the middle of the fifth course of the Moon, he reaches the point where he can be engaged *unconsciously* on the physical body. Through this activity in the dullness of his consciousness he creates for himself the first germinal predisposition to what is called "spirit self" (*Manas*). This "spirit self" attains its full unfolding in the course of the subsequent development of mankind. In its union with *Atma*, the "spirit-man," and with *Buddhi*, the "life-spirit," it is what later forms the higher, spiritual part of man. As on Saturn the *Thrones* or Spirits of Will permeated the "spirit-man" (*Atma*), and as on the Sun the *Cherubim* permeated the life-spirit (*Buddhi*) with wisdom, so now the *Seraphim* accomplish this for the "spirit-self" (*Manas*). They permeate it, and thereby implant in it a capacity which at later stages of development—on the earth—becomes that conceptualizing faculty of man by means of which, as a *thinking* being, he can enter into a relation with the world which surrounds him.

From the middle of the sixth course of the Moon onward, the "life-spirit" (*Buddhi*), from the middle of the seventh onward, the "spirit-man" (*Atma*) appear again, and these unite with the "spirit-self," so that at the end of the whole Moon era the "higher man" has been prepared. Then, together with all else that has developed on the Moon, the latter sleeps through a period of rest (*Pralaya*), in order to continue the course of his development on the earth planet.

While from the middle of the fifth Moon cycle onward into the sixth, man is working on his physical
body in dullness, the Spirits of Twilight are engaged on
his ether body. As has been shown, through their work
on the physical body in the preceding epoch (round),
they have now prepared themselves for relieving the Fire
Spirits in the ether body, who in turn take over from
the Spirits of Personality the work on the astral body.
At this time, these Spirits of Personality have ascended
to higher spheres.

The work of the Spirits of Twilight on the ether body
means that they connect their own states of consciousness
with the images of the consciousness of the ether body.
They thereby implant in these images the *joy* and the
pain which are caused by things. On the Sun the scene
of their corresponding activity had still been the merely
physical body. Hence joy and pain were there connected
only with the functions of this body and with *its* conditions. Now this becomes different. Joy and pain now become attached to the symbols which arise in the ether
body. In the dim human consciousness the Spirits of
Twilight thus experience a world of emotions. This is
the same world of emotions which man will experience
for himself in his earth consciousness.

At the same time, the Fire Spirits are active in the
astral body. They enable it to carry on an active perception and feeling of the environment. Joy and pain,
such as have been produced in the ether body by the
Spirits of Twilight in the manner just described, have an
inactive (passive) character; they present themselves as in-

active mirrorings of the outside world. But what the Fire Spirits produce in the astral body are vivid *emotions,* love and hate, rage, fear, horror, stormy passions, instincts, impulses and so forth. Because the Spirits of Personality (the *Asuras*) have previously inoculated this astral body with their nature, these emotions now appear with the character of selfhood, of separateness. One must now represent to oneself how at that time the ancestor of man is constituted on the Moon. He has a physical body through which in dullness he develops a "spirit self" (*Manas*). He has an ether body, through which the Twilight Spirits feel joy and pain; and finally he possesses an astral body which, through the Fire Spirits, is moved by impulses, emotions, and passions. But these three members of the Moon man still completely lack the object consciousness. In the astral body images flow and ebb, and in these there glow the emotions named above. When the thinking object consciousness will make its appearance on the earth, this astral body will be the subordinate carrier or the instrument of conceptual thinking. Now however, it unfolds in its own entire independence on the Moon. In itself it is more active here, more agitated than later on the earth. If one wishes to characterize it, one can say that it is an animal man. As such, it is on a higher level than the present-day animals of earth. It possesses the qualities of animality in a more complete way. In a certain respect these are more savage and unbridled than present-day animal qualities. Therefore, at this stage of his existence, one can call man a being which in its development stands midway between present-day

animals and man. If man had continued to advance in a
straight line along this path of development, he would
have become a wild, unrestrained being. The development of earth represents a toning down, a taming of the
animal character in man. This is caused by the thinking
consciousness.

If, as he had developed on the Sun, man was called
plant man, the man of the Moon can be called *animal
man*. That the latter can develop presupposes that the
environment also changes. It has been shown that the
plant-man of the Sun could only develop because an independent mineral realm was established alongside the
realm of this plant man. During the first two Moon eras
(rounds) these two earlier realms, plant realm and mineral realm, again emerge from the darkness. They are
changed only in that they both have become somewhat
coarser and denser. During the third Moon era a part
of the plant realm splits off. It does not take part in
the transition to coarseness. It thereby provides the substance out of which the animal nature of man can be
formed. It is this animal nature which, in its union with
the more highly formed ether body and with the newly
developed astral body, produces the threefold nature of
man which we have described above. The entire plant
world which had been formed on the Sun could not
develop into animality. For animals require the plant
for their existence. A plant world is the basis of an
animal world. As the Sun man could only elevate himself
into a plant by thrusting a portion of his companions
down into a coarser mineral realm, so this is now the case

with the animal man of the Moon. A portion of the beings which on the Sun still had the same plant nature as himself, he leaves behind him on the level of coarser plantlikeness. As the animal man of the Moon is not like the animals of today, but rather stands midway between present animal and present man, so too the mineral of the Moon lies between the mineral of today and the plant of today. The mineral of the Moon is something plantlike. The Moon rocks are not stones in the sense of today; they have an animated, sprouting, growing character. Similarly, the Moon plant has a certain character of animality.

The animal man of the Moon does not yet have firm bones. His skeleton is still cartilaginous. His whole nature is soft, compared to that of today. Hence his mobility too is different. His locomotion is not a walking, but rather a leaping, even a floating. This could be the case because the Moon of that time did not have a thin, airy atmosphere like that of present-day earth, but its envelope was considerably thicker, even denser than the water of today. He moved forward and backward, up and down in this viscous element. In this element also lived the minerals and animals from which he absorbed his nourishment. In this element was even contained the power which later on the earth was wholly transferred to the beings themselves—the power of fertilization. At that time man was not yet developed in the form of two sexes, but only in one. He was made out of his water air. But as everything in the world exists in transitional stages, in the last Moon periods, two-sexedness was already developing in a few animal man beings as a preparation for the later condition on the earth.

The sixth and seventh Moon cycles represent a kind of ebbing of all the processes we have described, but also the development of a kind of over-ripe condition, until the whole enters the period of rest (*Pralaya*) in order to pass in sleep into the existence of earth.

The development of the human astral body is connected with a certain cosmic process which must also be described here. When, after the period of rest which succeeds the cosmic era of the Sun, the latter again awakes and emerges from the darkness, then everything which lives on the thus developing planet still inhabits it as a whole. But this re-awakening Sun is nevertheless different from what it was before. Its substance is no longer luminous through and through, as it was previously; rather it now has darker portions. These separate out of the homogeneous mass, as it were. From the second cycle (round) onward, these portions appear more and more as an independent member; the Sun body thereby becomes biscuit-like. It consists of two parts, a considerably larger and a smaller one, which however are still attached to one another by a connecting link. In the third cycle these two bodies become completely separated. Sun and Moon are now two bodies, and the latter moves around the former in a circular orbit. Together with the Moon, all of the beings whose development has been described here, leave the Sun. The development of the astral body alone takes place on the split-off Moon. The cosmic process which we have characterized is the pre-condition of the further development described above. As long as the beings belonging to man absorbed their forces from their own solar habitat, their development could not attain the stage we have described. In the fourth cycle (round) the Moon is an independent planet, and what has been described concerning that period takes place on this Moon planet.

*

Here again, we shall present the development of the Moon planet and of its beings in a clearly summarized form.

I. The Moon is that planet on which man develops the image consciousness with its symbolical character.

II. During the first two cycles (rounds) the Moon development of man is prepared through a kind of repetition of the Saturn and Sun processes.

III. In the third cycle the human astral body comes into being through an outpouring of the Spirits of Motion.

IV. Concurrently with this process the Moon splits off from the re-awakened unified Sun body and revolves around the rest of the Sun. The development of the beings connected with man now takes place on the Moon.

V. In the fourth cycle the Spirits of Twilight inhabit the human physical body and thereby elevate themselves to the level of humanity.

VI. The developing astral body is inoculated with independence by the Spirits of Personality (*Asuras*).

VII. In the fifth cycle man begins to work in dullness on his physical body. Thereby the "spirit self" (*Manas*) joins the already existing monad.

VIII. In the ether body of man a kind of joy and pain develop during the Moon existence, which have a passive character. In the astral body on the other hand develop the emotions of rage, hate, the instincts, passions, and so forth.

IX. The two former realms, the plant and the mineral realm, which are thrust down to a lower level, are now joined by the animal realm, in which man himself exists at this time.

*

Toward the end of the whole universal era the Moon approaches more and more closely to the Sun, and when the time of rest (*Pralaya*) begins, again the two have become united in a whole, which then passes through the stage of sleep in order to awaken in a new universal era, that of the earth.

IN THE PRECEDING CHAPTERS has been shown how the components were successively formed which make up the so-called "lower nature of man"—the physical body, the ether body and the astral body. It has also been described how, with the appearance of a new body, the old ones must always be transformed so that they can become carriers and instruments of the one formed later. An advance of human consciousness is also associated with this progress. As long as the lower man has only a physical body, he possess merely an utterly dull consciousness, which is not equivalent even to that of dreamless sleep of the present, although for man of to-day this latter state of consciousness is in fact an "unconscious" one. In the time when the ether body appears, man reaches the consciousness which is his today in dreamless sleep. With the formation of the astral body a dim image consciousness makes its appearance, similar to, but not identical with the one man at present ascribes to himself while he is dreaming. The fourth, the current

condition of consciousness of earth man, will now be
described.

This present condition of consciousness develops in the
fourth great universal era, that of earth, which follows
the preceding Saturn, Sun, and Moon eras.

On Saturn the physical body of man was developed
in several stages. At that time it could not have been the
carrier of an ether body. And the latter was added only
during the course of the Sun. Simultaneously, the physical body was so transformed in the successive Sun cycles
that it could become the carrier of this ether body, or
in other words, that the ether body could work in the
physical body. During the Moon development the astral
body was added, and again the physical body and the
ether body were transformed in such a way that they
could provide suitable carriers and instruments for the
then appearing astral body. Thus, on the Moon, man is
a being composed of physical body, ether body, and
astral body. Through the ether body he is enabled to
feel joy and pain; through the astral body he is a being
with emotions, rage, hate, love and so forth.

As has been shown, higher spirits actively work on the
different members of his being. On the Moon the ether
body received the capacity for joy and pain through the
Spirits of Twilight; the emotions were implanted in the
astral body by the Fire Spirits.

At the same time, something else was taking place during the three great cycles on Saturn, Sun, and Moon.
During the last Saturn cycle the spirit man (*Atma*) was

formed with the help of the Spirits of Will (*Thrones*).
During the penultimate Sun cycle, the life-spirit (*Buddhi*)
was joined to it with the assistance of the Cherubim.
During the third from the last Moon cycle, the spirit-self
(*Manas*) united with the two others through the help of
the Seraphim. Thus actually two origins of man were
formed during these three great cycles: a lower man,
consisting of physical body, ether body, and astral body,
and a higher man, consisting of spirit man (*Atma*), life-
spirit (*Buddhi*), and spirit-self (*Manas*). The lower and
the higher nature of man followed separate paths at first.

The earth development serves to bring the two separate
origins of man together.

But first, after the seventh small cycle, all of Moon existence enters a kind of sleeping state (*Pralaya*). Thereby
everything becomes mixed together, so to speak, in a
homogeneous mass. The Sun and the Moon too, which
were separate in the last great cycle, again become fused
during the last Moon cycles.

When everything again emerges from the sleeping state
there must first be repeated in their essentials the Saturn
condition during a first small cycle, the Sun condition
during a second, and the Moon cycle during a third.
During this third cycle the beings on the Moon, which
has again been split off from the Sun, resume approximately the same forms of existence which they already
had on the Moon. There the lower man is a being intermediate between man of today and an animal; the plants
stand midway between the animal and plant natures of

today, and the minerals only half bear their lifeless character of today, while for the rest they are still half plants.

During the second half of this third cycle something else is already in preparation. The minerals harden, the plants gradually lose the animal character of their sensibility, and out of the uniform species of animal man there develop two classes. One of these remains on the level of animality, while the other is subjected to a division of the astral body into two parts. The astral body splits into a lower part, which continues to be the carrier of the emotions, and a higher part, which attains to a certain independence, so that it can exercise a kind of mastery over the lower members, over the physical body, the ether body, and the lower astral body. Now the Spirits of Personality seize upon this higher astral body and implant in it just that independence we have mentioned, and therewith also selfishness. Only in the lower human astral body do the Fire Spirits now accomplish their work, while in the ether body the Spirits of Twilight are active, and in the physical body that power entity begins its work which one can describe as the real ancestor of man. It is the same power entity which formed the spirit man (*Atma*) with the help of the Thrones on Saturn, the life-spirit (*Buddhi*) with the assistance of the Cherubim on the Sun, and the spirit-self (*Manas*) together with the Seraphim on the Moon.

But now this changes. Thrones, Cherubim, and Seraphim ascend to higher spheres, and the higher man now receives the assistance of the Spirits of Wisdom, of Motion, and of Form. These are now united with spirit-self, life-spirit, and spirit man (with *Manas—Buddhi—Atma*). With the assistance of these entities the human power being characterized above develops its physical body during the second half of the third earth cycle. It is the Spirits of Form which act here in the most significant way. They already form the human physical body so that it becomes a kind of precursor of the later human body of the fourth cycle (the present one, or the fourth round).

In the astral body of the animal beings which have been left behind, it is exclusively the Fire Spirits which remain active, while in the ether body of the plants it is the Spirits of Twilight. On the other hand, the Spirits of Form participate in the transformation of the mineral realm. They make the latter hard, that is, implant rigid and fixed forms in it.

One must not imagine, however, that the sphere of activity of the spirits we have mentioned was confined only to what has been characterized. Always it is only the main directions of their activities which are meant. In a subordinate way all the spirit beings participate everywhere. Thus at the time indicated, the Spirits of Form also have certain functions to perform in the physical plant and animal bodies, and so forth.

After all this has taken place, around the end of the third earth cycle all the entities—including Sun and Moon —again become fused and then pass through a shorter stage of sleep (a small *Pralaya*). At that time everything

is again a uniform mass (a chaos), and at the end of this
stage the fourth earth cycle begins in which we are at
present.

At first, everything which already previously had had
a being in the mineral, plant, animal, and human realms,
begins to separate out of the uniform mass in germinal
form. First there can re-emerge as *independent* germs
only those human ancestors on whose higher astral bodies
the Spirits of Personality have worked in the preceding
small cycle. All other beings of the mineral, plant, and
animal realm do not yet lead an independent existence
here. At this stage everything is still in that high spiritual
condition which is called the "formless" or *Arupa* condition. At the present stage of development, only the
highest human thoughts—for example, mathematical and
moral ideals—are woven of that substance which was
proper to all beings at the stage we are describing. That
which is lower than these human ancestors can only appear as an activity in a higher being. The animals exist
only as states of consciousness of the Spirits of Fire, the
plants as states of consciousness of the Spirits of Twilight.
The minerals have a double existence in thought. First
they exist as thought germs in the human ancestors mentioned above, then as thoughts in the consciousness of the
Spirits of Form. The "higher man" (spirit man, life-
spirit, spirit-self) also exists in the consciousness of the
Spirits of Form.

By degrees a densification of everything now occurs.
But at the next stage, the density as yet does not exceed
the density of thoughts. At this stage, however, the animal

beings which originated in the preceding cycle, can emerge. They separate out of the consciousness of the Fire Spirits and become independent thought beings. This stage is called that of the "formed" or *Rupa* condition. Man advances in it insofar as his previously formless, independent thought body is clothed by the Spirits of Form in a body of coarser, formed thought substance. The animals as independent beings, here consist exclusively of this substance.

Now a further densification takes place. The condition which is now attained can be compared with that out of which the conceptions of the dreamlike image consciousness are woven. One calls this the "astral" stage.

The human ancestor again advances. In addition to the other two components, his being receives a body which consists of the substance just characterized. He now has the inner formless core of being, a thought body, and an astral body. The animals receive a similar astral body, and the plants emerge from the consciousness of the Spirits of Twilight as independent astral entities.

In the further course of the development, the densification now advances to that condition which is called the physical. At first we deal with the most refined physical condition, with that of the most refined ether. From the Spirits of Form the human ancestor receives the most refined ether body as an addition to his earlier components. He consists of a formless thought core, a formed thought body, an astral body, and an ether body. The animals have a formed thought body, an astral body, and an ether body; the plants have an astral and an ether body; the minerals first emerge here as independent ether forms. At this stage of development we are concerned with four realms: a mineral, a plant, an animal, and a human realm. Along with these however, in the course of the development up to this point three other realms have come into existence. In the time when the animals separated from the Fire Spirits at the thought stage (*Rupa* stage), the Spirits of Personality also separated certain entities out of themselves. These consist of indefinite thought substance which gathers together, dissolves in a cloudlike manner, and thus flows along. One cannot speak of them as of independent entities, but only of an irregular, general mass. This is the first elementary realm. At the astral stage something similar separates from the Fire Spirits. It consists of shadowy images or phantoms similar to the conceptions of the dreamlike image consciousness. They form the second elementary realm. In the beginning of the physical stage, indefinite image-like entities finally separate out of the Spirits of Twilight. They too have no independence, but they can manifest forces which are similar to the passions and emotions of men and animals. These non-independent, buzzing emotions form the third elementary realm. For beings which are endowed with a dreamlike image consciousness, or with conscious image consciousness, these creations of the third elementary realm are perceptible as a flooding light, as flakes of color, as smell, taste, as various tones and sounds. But all such perceptions must be imagined to be phantom-like.

One must therefore imagine the earth, when it condenses as a refined etheric body out of its astral precursor,

to be a conglomerate of an etheric mineral basic mass
and of etheric plant, animal, and human beings. Filling
the interstices as it were, and also permeating the other
beings, are the creatures of the three elementary realms.

This earth is inhabited by the higher spiritual entities,
which, in the most diverse ways, are active in the realms
we have mentioned. They form a spirit community, so
to speak, a spirit state, and their dwelling-place and workshop is the earth, which they carry with them as a snail
does its shell. In all this it must be borne in mind that
what today is separated from the earth as Sun and Moon,
is still entirely united with the earth. Only later do
both heavenly bodies separate from the earth.

The "higher man" (spirit man—life-spirit—spirit-self,
Atma—Buddhi—Manas) has as yet no independence at
this stage. He still constitutes a member of the spirit
state, and for the time being is bound to the Spirits of
Form, as a human hand is bound to a human organism
as a dependent member.

With this we have followed the formation of the earth
to the beginning of its physical condition. In what follows, we shall show how everything in this condition
develops further. The previous development will then
pass over into what has already been said in preceding
chapters of the Akasha Chronicle about the progress of
earth.

States of development such as those which have been
mentioned here as a formless, a formed, an astral, and a
physical condition, which thus constitute differences in a
smaller cycle (a round), are called 'globes' in theosophi-

cal texts. In this respect one therefore speaks of an *Arupa,*
a *Rupa,* an astral and a physical globe. Some have considered such a designation incorrect. But here we shall
not speak further of matters of nomenclature. Indeed, it
is not names which are important, but rather the things
themselves. It is better to endeavor to describe the latter
as well as possible than to worry very much about names.
These must after all *always* be incorrect in a certain sense.
For to facts of the spiritual world one must give names
which have come from the world of the senses, and
therefore one can speak only by way of similes.

*

The description of the development of the world of
man has been brought to the point where the earth
reaches the beginning of its physical densification. One
should now represent to oneself the condition of development of this world of man at this stage. What later appears as sun, moon, and earth is still united in a single
body. This body possesses only refined etheric matter. It
is only within this matter that the beings which later
will appear as men, animals, plants, and minerals have
their existence. For the further progress of the development, the one heavenly body must first separate into two,
of which one becomes the later sun, while the other
contains the later earth and the later moon in a still
united form. Only later does a process of division also
take place in this latter heavenly body; that which becomes the present-day moon is extruded, and the earth

alone remains as the dwelling-place of man and of his fellow-creatures.

The student of ordinary theosophical literature should understand clearly that the separation of the *one* heavenly body into two took place in the period in which this literature places the development of the so-called second principal human race. The human ancestors of this race are described as forms with refined etheric bodies. But one must not imagine that they could have developed on our present earth after it had already separated itself from the sun and had expelled the moon. After this separation, such etheric bodies were no longer possible.

If one follows the development of mankind in that cycle to which our description has now come, and which leads us into the present, one becomes aware of a series of principal conditions, of which our current one is the fifth.

The preceding expositions from the Akasha Chronicle have already dealt with these conditions. Here we shall only repeat what is necessary for a further deepening of the discussion.

The first principal condition shows the human ancestors as highly refined etheric entities. The usual theosophical literature somewhat inexactly calls these entities the first principal race. This condition essentially continues during the second epoch, in which this literature places the second principal race. Until this stage of development sun, moon, and earth are still one heavenly body. Now the sun splits off as an independent body. It thereby takes from the earth, which is still united with the moon, all those forces through which the human ancestors could be maintained in their etheric condition. With the splitting-off of the sun a densification of the human forms and also of the forms of man's fellow-creatures takes place. These creatures must now adapt themselves to their new dwelling-place, as it were.

But it is by no means only the material forces which are taken away from this dwelling-place. Spiritual entities, of whom it has been said that they formed a spirit community on the *one* heavenly body we have described, also leave at the same time. Their existence remains more intimately connected with the sun than with the heavenly body which the sun has thrust out from itself. If these entities had remained united with the forces which later develop on the earth and on the moon, they themselves could not have developed further to their appropriate levels. They needed a new dwelling-place for this further development. This is provided for them by the sun after it—so to speak—has cleansed itself of the earth and moon forces. At the stage at which these beings now are, they can act on earth and moon forces only from the outside, from the sun.

One can see the reason for the separation we have described. Until this time, certain entities which are higher than man have gone through their development on the *one* heavenly body characterized above; they now lay claim to a part of it for themselves, and leave the rest to man and his fellow-creatures.

The consequence of the splitting-off of the sun was a radical revolution in the development of man and of his fellow-creatures. They fell as it were from a higher level of existence to a lower. They had to do this, for they lost the immediate connection with those higher beings. They would have entered entirely into a blind alley in their own development if other universal events had not taken place, through which progress was stimulated anew and the development directed into quite different channels.

With the forces which at present are united in the split-off moon, and which at that time were still within the earth, further progress would have been impossible. Present-day mankind could not have been produced with these forces, but, instead, only a kind of being in which the emotions of rage, hate and so forth, developed during the third great cycle, the Moon existence, would have increased to the point of the immoderate and animal.

During a certain period, this was in fact the case. The immediate consequence of the splitting-off of the sun was the arising of the third principal condition of the ancestors of man, which in theosophical literature is designated as the third principal race, the Lemurian. Again, the designation "race" for this condition of development is not an especially fortunate one. For in a real sense, the human ancestors of that time cannot be compared with what today one designates as "race." One must be completely clear about the fact that the evolutionary forms of the distant past as well as of the future are so entirely different from those of today that our present appellations can only serve as makeshifts, and really lose all meaning in relation to these remote epochs.

Actually, one can only begin to speak of "races" in connection with the development attained in about the second
third of the third principal condition identified above
(the Lemurian). Only then is formed what today one
calls "races." This "racial character" is retained in the
period of the Atlantean development, and further into
our time of the fifth principal condition. But already at
the end of our fifth era, the word "race" will again lose
all sense. In future, mankind will be divided into parts
which it will be impossible to designate as "races." In
this respect, ordinary theosophical literature has caused
much confusion. This has especially been done by Sin-
nett's *Esoteric Buddhism,* the book which, on the other
hand, has the great merit of having been the first to
popularize the theosophical world-outlook in recent times.
In this book the universal development is described as
if, throughout the cosmic cycles, the "races" forever repeated themselves in the same way. But this is by no
means the case. That which deserves to be called "race"
also *comes into being* and *perishes.* One should only
use the expression "race" for a certain span in the development of mankind. Before and after this, there are evolutionary forms which are something totally different from
"races."

Only because the true deciphering of the Akasha Chronicle fully authorizes one to make such a remark have we
presumed to make it here. In this the decipherer knows
himself to be in complete accord with true occult spiritual investigation. Otherwise it could never occur to him
to make such objections against the meritorious books

of theosophical literature. He might also make the really quite superfluous remark that the inspirations of the great teachers mentioned in *Esoteric Buddhism* are not contradicted by what has been described here, but that the misunderstanding has only been produced by the fact that the author of that book has transposed the wisdom of these inspirations, which is difficult to express, into modern every-day human language, in his own way.

The third principal condition of the development of mankind presents itself as the one in which the "races" first came into being. This event was brought about by the separation of the moon from the earth. This separation was accompanied by the originating of the two sexes. This stage of the development of mankind has been repeatedly referred to in the descriptions from the "Akasha Chronicle." When the earth, still united with the moon, split off from the sun, a male and a female sex did not as yet exist within mankind. Each human being combined the two sexes within its still highly refined body.

It must be remembered however that these double-sexed human ancestors were on a low level of development as compared with present-day man. The lower impulses acted with immeasurable energy, and nothing of a spiritual development as yet existed. That the latter was stimulated and that thereby the lower impulses were confined within certain bounds, is connected with the fact that, at the same time at which earth and moon separated, the former came into the sphere of influence of other heavenly bodies. This extremely significant co-operation of the earth with other heavenly bodies, its meeting with foreign planets,

in the time which theosophical literature calls the Lemurian, will be related in a further chapter of the "Akasha
Chronicle."

The same course of development will be described once
more, but from a different point of view. This is done for
a quite definite reason. One can never look at the truths
about the higher worlds from too many aspects. One
should realize that from any one aspect it is possible to
give only the poorest sketch. And when one looks at the
same thing from the most diverse aspects, the impressions
one receives in this way only gradually complement each
other to form an ever more animated *picture*. Only such
pictures, not dry, schematic concepts, can help the man
who wants to penetrate into the higher worlds. The more
animated and colorful the pictures, the more can one hope
to approach the higher reality.

It is obvious that it is just the pictures from the higher
worlds which arouse mistrust in many today. A person is
quite content to be given conceptual schemes and classifi-
cations—with as many names as possible—of *Devachan*, of
the development of the planets, and so forth; but he becomes more difficult when somebody presumes to describe
the supersensible worlds as a traveller describes the landscapes of South America. Yet one should realize that it is
only through fresh, animated pictures that one is given
something useful, not through dead schemes and names.

IN THIS DESCRIPTION, we shall take man as our point of departure. As he lives on the earth, man at present consists of the physical body, the ether or life body, the astral body and the "I." This fourfold human nature has in itself the dispositions for a higher development. The "I" by its own initiative transforms the "lower" bodies, and thereby incorporates into them higher parts of human nature. The ennobling and purifying of the astral body by the "I" causes the development of the "spirit self" (*Manas*), the transformation of the ether or life body creates the life spirit (*Buddhi*), and the transformation of the physical body creates the true "spirit man" (*Atma*). The transformation of the astral body is in full progress in the present period of the development of the earth; the conscious transformation of the ether body and of the physical body belongs to later times; at present it has begun only among the initiates—those trained in the science of the spirit and their pupils.

This threefold transformation of man is a conscious one; it was preceded by one more or less unconscious during the previous development of the earth. It is in this unconscious transformation of astral body, ether body, and physical body that one must seek the origin of the sentient soul, of the intellectual soul, and of the consciousness soul.

One must now make clear to oneself which one of the three bodies of man (the physical, the ether, and the astral body) is in its way most perfect. One can easily be tempted to consider the physical body as the lowest and therefore as the least perfect. However, in this one would be in error. It is true that the astral body and the ether body will attain a high degree of perfection in the future, but at present the physical body is more perfect *in its way* than are they in theirs. Only because man has this physical body in common with the lowest natural realm on earth, the mineral realm, is it possible for the error we have mentioned to arise. For man has the ether body in common with the higher plant realm, and the astral body with the animal realm.

Now it is true that the physical body of man is composed of the same substances and forces which exist in the wider mineral realm, but the manner in which these substances and forces interact in the human body is the expression of wisdom and perfection in the structure. One will soon convince himself of the truth of this statement if he undertakes to study this structure not merely with the dry intellect but with his whole feeling soul. One can take any part of the human physical body as the subject for this contemplation, for instance the highest part of the upper thigh bone. This is not an amorphous massing of substance, but rather is joined together in the most artful manner, out of diminutive beams which run in different directions. No modern engineering skill could fit a bridge or something similar together with such wisdom. Today such things are still beyond the reach of the most perfect human wisdom. The bone is constructed in this wise fashion so that, through the arrangement of the small beams, the necessary carrying capacity for the support of the human torso can be attained with the least amount of substance. The least amount of matter is used in order to achieve the greatest possible effect in terms of force. In face of such a "masterwork of natural architecture," one can only become lost in admiration. No less can one admire the miraculous structure of the human brain or heart, or of the totality of the human physical body. One should compare with it the degree of perfection which for example the astral body has attained at the present stage of development of mankind. The astral body is the carrier of pleasure and distaste, of the passions, impulses and desires and so forth. But what attacks this astral body perpetrates against the wise arrangement of the physical body! A large part of the stimulants which man consumes are poisons for the heart. From this it can be seen that the activity which produces the physical structure of the heart proceeds in a wiser manner than the activity of the astral body, which even runs counter to this wisdom. It is true that in future the astral body will advance to higher wisdom; at present, however, it is not as perfect *in its way* as is the physical body. Something similar could be shown to be true for the ether body, and also for the "I," that being which, from moment to moment, must struggle gropingly toward wisdom through error and illusion.

If one compares the levels of perfection of the parts of the human being, one will easily discover that at present the physical body is in its way the most perfect, that the ether body is less perfect, the astral body still less, and that in its way the least perfect part of man at present is the "I." This is due to the fact that in the course of the planetary development of the human dwelling-place the physical body of man has been worked on the longest. What man today carries as his physical body has lived through all the developmental stages of Saturn, Sun, Moon, and earth, up to the present stage of the latter. All the forces of these planetary bodies have successively worked on this body, so that gradually it has been able to attain its present degree of perfection. It is thus the *oldest* part of the present-day human entity.

The ether body, as it now appears in man, did not exist at all during the Saturn period. It was only added during the Sun development. Hence the forces of four planetary bodes have not worked on it as on the physical body, but only those of three, namely, of Sun, Moon, and earth. Therefore only in a future period of development can it become as perfect in its way as is the physical body at present. The astral body joined the physical body and the ether body only during the Moon period, and the "I" did so only during the earth period.

One must represent to oneself that the physical human body attained a certain stage of its development on Saturn, and that this development was carried forward on the Sun in such a way that from that time on the physical body could become the carrier of an ether body. On Saturn this physical body had attained a point where it was an extremely complex mechanism, which however had nothing in it of life. The complicatedness of its structure finally caused it to disintegrate. For this complicatedness had reached such a degree that this physical body could no longer maintain itself by means of the merely mineral forces which were acting in it. It was through this collapse of the human bodies that the decline of Saturn was brought about.

Of the present natural realms, namely the mineral realm, the plant realm, the animal realm, and the human realm, Saturn had only the last-named. What one knows today as animals, plants, and minerals did not yet exist on Saturn. Of the present four natural realms, there existed on this heavenly body only man in his physical body, and this physical body was in fact a kind of complicated mineral. The other realms came into existence because not all beings could attain full development on the successive heavenly bodies. Thus only a part of the human bodies developed on Saturn attained the full Saturn goal. Those human bodies which did attain this goal were awakened, so to speak, to a new existence in their old form during the Sun period, and this form was permeated with the ether body. They thereby developed to a higher level of perfection. They became a kind of plant men. That portion of the human bodies however which had not been able to attain the full goal of development

on Saturn had to continue during the Sun period what
they had previously not completed, but under considerably more unfavorable conditions than those which existed for *this* development on Saturn. They therefore
fell behind the portion which had attained the full goal
on Saturn. Thus on the Sun another natural realm came
into being in addition to the human realm.

It would be erroneous to believe that all the organs
in the present-day human body already began to be developed on Saturn. This is not the case. Rather it is par-
ticularly the sensory organs in the human body which
have their origin in this ancient time. It is the first rudiments of eyes, ears and so forth which have such an
early origin, rudiments which formed on Saturn in somewhat the same way that "lifeless crystals" now form on
earth; the corresponding organs then attained their present
form by again and again transforming themselves in the
direction of greater perfection in each of the succeeding
planetary periods. On Saturn they were physical instruments, and nothing else. On the Sun they were transformed, because an ether or life body permeated them.
They were thereby brought into the life process. They
became *animated* physical instruments. To them were
added those parts of the human physical body which cannot develop at all except under the influence of an ether
body: the organs of growth, of nourishment, and of reproduction. Of course the first rudiments of these organs,
as they developed on the Sun, again do not resemble in
perfection the form which they have at present.

The highest organs which the human body at that

time acquired through the interaction of physical body
and ether body were those which at present have developed into the *glands*. The physical human body on the
Sun is thus a system of glands, on which sensory organs
of a corresponding level of development are impressed.

The development continued on the Moon. To the
physical body and the ether body is added the astral
body. Thereby the first rudiment of a nervous system
is integrated into the glandular sensory body. One can
see that the physical human body becomes more and
more complicated in the successive planetary development
periods. On the Moon it is composed of nerves, glands,
and senses. The senses have behind them a two-fold process of transformation and perfection, while the nerves are
at their first stage. If one looks at the Moon man as a
whole, he consists of three parts: a physical body, an ether
body, and an astral body. The physical body is tripartite;
its partition is the result of the work of the Saturn, Sun,
and Moon forces. The ether body is only bipartite. It has
in itself only the effect of the work of Sun and Moon,
and the astral body is still unipartite. Only the Moon
forces have worked on it.

Through the absorption of the astral body on the Moon,
man has become capable of a life of sensation, of a certain inwardness. Within his astral body he can form
images of what takes place in his environment. These
images in a certain respect are to be compared with the
dream images of present-day human consciousness, but
they are more vivid and colorful, and, most important,
they relate to events in the outside world, while present-

day dream images are mere echoes of daily life or are
otherwise *unclear* mirrorings of inner or outer events.
The images of the Moon consciousness corresponded completely to whatever they were related to externally. Assume for instance that a Moon man as he has just been
characterized, consisting of physical body, ether body, and
astral body, had approached another Moon being. It is
true that he could not have perceived the latter as a
spatial object, for this has become possible only in the
earth consciousness of man; but within his astral body
would have arisen an image which in its color and shape
would have quite exactly expressed whether the other
being was well or ill disposed toward this Moon man,
whether it would be useful or dangerous to him. As a
result, the Moon man could regulate his behavior entirely in accordance with the images which arose in his
image consciousness. These images were a complete means
of orientation for him. The physical instrument which
the astral body needed in order to enter into relation
with the lower natural realms was the nervous system,
integrated into the physical body.

In order that the transformation of man described
here could occur during the Moon period, the assistance
of a great universal event was needed. The integration
of the astral body and the related development of a
nervous system in the physical body was only made possible by the fact that what had previously been *one* body,
the *Sun,* split into *two*—into *Sun* and *Moon.* The former
advanced to the state of a fixed star, the latter remained
a planet—which the Sun also had been—and began to circle

around the Sun, from which it had split off. Through this a significant transformation took place in everything which lived on Sun and Moon. Here for the moment we shall follow this process of transformation only insofar as it concerns the life of the Moon. Man, consisting of physical body and astral body, had remained united with the Moon when it split off from the Sun. He thereby entered into entirely new conditions of existence. For the Moon took with it only a part of the forces contained in the Sun, and this part now acted on man from his own heavenly body; the Sun had retained the other part of the forces within itself. This latter part is now sent from the outside to the Moon and hence also to its inhabitant, man. If the previous relationship had remained in existence, if all the Sun forces had continued to reach man from his own scene of activity, that inner life which shows itself in the arising of the images of the astral body could not have developed. The Sun force continued its activity on the physical body and ether body *from the outside;* it had already acted on both of these previously. But it liberated a portion of these two bodies for influences which emanated from the Moon, the heavenly body newly created by a splitting-off. Thus, on the Moon, man was under a two-fold influence, that of the Sun and that of the Moon. It is to be ascribed to the influence of the Moon that out of the physical and the ether body there developed those parts which permitted the imprinting of the astral body. An astral body can only create images when the Sun forces reach it from outside rather than from its own planet. The Moon influences transformed the sensory rudiments and the glandular organs in such a way that a nervous system could be integrated into them; and the Sun influences brought it about that the images for which this nervous system was the instrument corresponded to the external Moon events in the manner described above.

The development could only progress to a certain point in this manner. Had this point been overstepped, the Moon man would have become hardened in his inner life of images, and thereby he would have lost all connection with the Sun. When the time had come, the Sun again absorbed the Moon, so that again for some time both were *one* body. The union lasted until man was far enough advanced so that his hardening, which would have had to take place on the Moon, could be prevented by a new stage of development. When this had occurred a new separation took place, but this time the Moon took Sun forces along with itself which previously it had not received. Through this it came about that another separation took place after some time. What had last split off from the Sun was a heavenly body which contained all the forces and beings at present living on earth and moon. Thus the earth still contained within itself the moon which now circles around it. If the latter had remained within it, it could never have become the scene of any human development, including the present one. The forces of the present moon first had to be cast off, and man had to remain on the thus purified earthly scene and continue his development there. In this way three heavenly bodies developed out of the old Sun. The forces

of two of these heavenly bodies, the new sun and the
new moon, are sent to the earth and hence to its inhabitant from the outside.

Through this progress in the development of the
heavenly bodies it became possible that into the tripartite
human nature, as it still had been on the Moon, the
fourth part, the "I," integrated itself. This integration
was connected with a perfecting of the physical body,
the ether body, and the astral body. The perfecting of
the physical body consisted in that the system of the
heart was incorporated in it as the preparer of *warm
blood*. Of course, now the sensory system, the glandular
system, and the nervous system had to be transformed
in such a way that in the human organism they would
be compatible with the newly added system of the warm
blood. The sensory organs were so transformed that out
of the mere image consciousness of the old Moon the
object consciousness could develop, which makes possible
the perception of *external* objects, and which man at
present possesses from the time he awakes in the morning until he falls asleep in the evening. On the old Moon
the senses were not yet open to the outside; the images
of consciousness arose from within, and just this opening
of the senses to the external is the achievement of the
earth development.

It has been stated above that not all of the human
bodies formed on Saturn attained the goal which was
set for them there, and that on the Sun, alongside the
human realm in its form of that time, a second natural
realm developed. One must realize that at each of the

subsequent stages of development, on Sun, Moon, and earth, there were always beings which fell short of their goals and that through this the lower natural realms came into existence. The animal realm, which is closest to man, had already fallen behind on Saturn, but partially made up the development under unfavorable conditions on Sun and Moon, so that while on the earth it was not as far advanced as man, in part it still had the capacity to receive warm blood as he did. For warm blood existed in *none* of the natural realms before the period of earth. The present-day cold-blooded (or variably warm) animals and certain plants came into existence because certain beings of the lower Sun realm again fell short of the stage which the other beings of this realm attained. The present-day mineral realm came into existence last, in fact only during the earth period.

The fourfold man of earth receives from sun and moon the influences of those forces which have remained connected with these heavenly bodies. From the sun those forces reach him which further progress, growth, and becoming; from the moon come the hardening, forming forces. If man stood only under the influence of the sun he would dissolve in an immeasurably rapid process of growth. It is for this reason that formerly, after an appropriate time, he had to leave the Sun and to receive a retarding of his over-rapid progress on the split-off old Moon. But if then he had remained permanently connected with the latter, this retarding of his growth would have hardened him in a rigid form. Therefore he advanced to the development of earth, within which the

two influences counterbalance each other in an appropriate way. At the same time the point is reached where
something higher—the soul—is integrated as an inner
entity within the quadripartite human being.

In its form, in its activities, movements and so forth,
the physical body is the expression and the effect of what
takes place in the other parts, in the ether body, the
astral body, and the "I." In the descriptions from the
"Akasha Chronicle" which we have given up to this point,
it has become apparent how, in the course of development, these other parts of the human entity gradually
intervened in the formation of the physical body. During
the Saturn development none of these other parts was as
yet associated with the physical human body. But the
first beginning of its development was made in that time.
However, one must not think that the forces which later
acted on the physical body from the ether body, the
astral body, and the "I" did not already act on it during
the Saturn period. They were already acting at that time,
but in a certain sense from the outside, not from within.
The other parts had not yet been formed, had not yet
been united with the physical human body as individual
entities; but the forces which were later united in them
acted as it were from the environment—the atmosphere—
of Saturn and formed the first beginning of this body.
This beginning was then transformed on the Sun because a part of those forces now formed the separate human ether body, and now acted on the physical body no
longer merely from the outside, but from the inside. The
same thing occurred on the Moon with respect to the

astral body. On the earth the physical body was transformed for the fourth time by becoming the dwelling-
place of the "I," which now works within it.

One can see that, to the eye of the scientist of the spirit,
the physical body is not something fixed, something per-
manent in its form and manner of acting. It is undergoing
a constant process of transformation. And such a transformation is also taking place in the current earth period
of the body's development. One can only understand
human life if one is in a position to form a conception
of this transformation.

A consideration of the human organs from the point
of view of the science of the spirit shows that these are
at very different stages of development. There are organs in the human body which, in their present form,
are in a descending, others which are in an ascending development. In future, the former will lose their importance for man more and more. The time of the flower-
ing of their functions is behind them; they will become
atrophied and finally disappear from the human body.
Other organs are in an ascending development; they
contain much which now is only present in a germinal
state, as it were; in future they will develop into more
perfect forms with a higher function. Among the former
organs belong, for instance, those which serve for reproduction, for the bringing into existence of like beings.
In future their function will pass to other organs and
they themselves will sink into insignificance. There will
come a time when they will be present on the human
body in an atrophied condition, and one will then have

to regard them only as evidences of the preceding development of man.

Other organs, as for instance the heart and neighboring formations, are at the beginning of their development in a certain respect. What now lies in them in a germinal state will reach its full flower only in the future. For in the conception of the science of the spirit, the heart and its relation to the so-called circulation of the blood are seen as something quite different from what contemporary physiology, which in this respect is completely dependent on mechanistic-materialistic concepts, sees in them. In so doing, this science of the spirit succeeds in casting light on facts which are well-known to contemporary science, but for which with the means at its disposal, the latter cannot give anything like a satisfactory explanation. Anatomy shows that in their structure the muscles of the human body are of two kinds. There are those whose smallest parts are smooth bands, and those whose smallest parts show a regular transverse striation. Now the smooth muscles in general are those which in their movements are independent of human volition. For instance, the smooth muscles of the intestine push the food pulp along in regular movements, upon which the human volition has no influence. Those muscles which are found in the iris of the eye are also smooth. These muscles bring about the movements through which the pupil of the eye is enlarged when the latter is exposed to a small amount of light, and contracted when much light flows into the eye. These movements too are independent of human volition. On the other hand, those

muscles are striated which mediate movements under the
influence of human volition, for example, the muscles
by which the arms and legs are moved. The heart, which
after all is also a muscle, constitutes an exception to this
general condition. In the present period of human development, the heart is not subject to volition in its movements, yet it is a "transversely striated" muscle. The
science of the spirit indicates the reason for this. The
heart will not always remain as it is now. In the future
it will have a quite different form and a changed function.
It is on the way to becoming a voluntary muscle. In the
future it will execute movements which will be the effects of the inner soul impulses of man. It already shows
what significance it will have in the future, when the
movements of the heart will be as much an expression of
the human will as the lifting of the hand or the advancing
of the foot are today.

This conception of the heart is connected with a comprehensive insight of the science of the spirit into the
relation of the heart to the so-called circulation of the
blood. The mechanical-materialistic doctrine of life sees
in the heart a kind of pumping mechanism which drives
the blood through the body in a regular manner. Here
the heart is the cause of the movement of the blood. The
insight of the science of the spirit shows something quite
different. For this insight, the pulsing of the blood, its
whole inner mobility, are the expression and the effect
of the processes of the soul. The soul is the cause of the
behavior of the blood. Turning pale through feelings of
fear, blushing under the influence of sensations of shame,

are coarse effects of processes of the soul in the blood. But everything which takes place in the blood is only the expression of what takes place in the life of the soul. However, the connection between the pulsation of the blood and the impulses of the soul is a deeply mysterious one. The movements of the heart are not the cause, but the consequence of the pulsation of the blood.

In the future, through voluntary movements, the heart will carry what takes place in the human soul into the external world.

Other organs which are in a similarly ascending development are the organs of respiration in their function as instruments of speech. At present by their means man can transform his thoughts into air waves. He thereby impresses upon the external world what he experiences within himself. He transforms his inner experiences into air waves. This wave motion of the air is a rendering of what takes place within him. In the future he will in this way give external form to more and more of his inner being. The final result in this direction will be that through his speech organs which have arrived at the height of their perfection, he will produce his own kind. Thus the speech organs at present contain within themselves the future organs of reproduction in a germinal state. The fact that mutation (change of voice) occurs in the male individual at the time of puberty is a consequence of the mysterious connection between the instruments of speech and reproduction.

The entire human physical body can be considered in this way from the point of view of the science of the spirit. It was only intended to give a few examples here. In the science of the spirit, both an anatomy and a physiology exist. The anatomy and physiology of the present will have to let themselves be fertilized by the anatomy and physiology of the science of the spirit in a not very distant future, and will even transform themselves completely into the latter.

In this area it becomes especially apparent that results such as those given above must not be built on mere inferences, on speculations such as conclusions by analogy, but must only proceed from the true research of the science of the spirit. This must necessarily be emphasized, for it happens only too easily that once they have gained some insights, zealous adherents of the science of the spirit continue to spin their ideas in empty air. It is no miracle when only phantasms are produced in this way, and, in fact, they do abound in these areas of research.

One could, for instance, proceed to draw the following conclusion from the description given above: Because the human organs of reproduction in their present form will in the future be the first to lose their importance, they therefore were the first to receive it in the past, hence they are in a sense the oldest organs of the human body. Just the contrary of this is true. They were the last to receive their present form and will be the first to lose it again.

The following presents itself to spiritual scientific research. On the Sun, the physical human body had in a certain respect moved up to the level of plant existence.

At that time it was permeated only by an ether body. On the Moon it took on the character of the animal body, because it was permeated by the astral body. But not all organs participated in this transformation into the animal character. A number of parts remained on the plant level. On the earth, after the integration of the "I," when the human body elevated itself to its present form, a number of parts still bore a decided plant character. But one must not imagine that these organs looked exactly like our present-day plants. The organs of reproduction belong among these organs. They still exhibited a plant character at the beginning of the earth development. This was known to the wisdom of the old Mysteries. The older art which has retained so much of the traditions of the Mysteries, represents hermaphrodites with plant-leaf-like organs of reproduction. These are precursors of man which still had the old kind of reproductive organs (which were double-sexed). For example, this can be seen clearly in a hermaphrodite figure in the Capitoline Collection in Rome. When one looks into these matters one will also understand for instance the true reason for the presence of the fig leaf on Eve. One will accept true explanations for many old representations, while contemporary interpretations are, after all, only the result of a thinking which is not carried to its conclusion. We shall only remark in passing that the hermaphrodite figure mentioned above shows still other plant appendages. When it was made, the tradition still existed that in a very remote past certain human organs changed from a plant to an animal character.

All these changes of the human body are only the expression of the forces of transformation which lie in the ether body, the astral body, and the "I." The transformations of the physical human body accompany the acts of the higher parts of man. One can therefore understand the structure and the activity of this human body only if one absorbs oneself in the "Akasha Chronicle," which shows how the higher changes of the more spiritual and mental parts of man take place. Everything physical and material finds its explanation through the spiritual. Light is shed even on the *future* of the physical if one studies the spiritual.

If we are to acquire new capacities through repeated incarnations in the successive races, if in addition nothing of what the soul has acquired through experience is again to be lost from its storehouse,—how is it to be explained that in mankind of today absolutely nothing remains of the capacities of the will, of the imagination, of the mastery of natural forces which were so highly developed in those periods?

It is a fact that of the capacities which the soul has acquired in its transition through a stage of development, nothing is lost. But when a new capacity is acquired, the one previously gained assumes a different form. It then no longer manifests itself in its own character, but as a *basis* for the new capacity. Among the Atlanteans for instance, it was the faculty of memory which was acquired. Contemporary man indeed can form only a very weak conception of what the memory of an Atlantean could accomplish. But all that appears as *innate* concepts in our fifth root race, in Atlantis was only acquired through the memory. The concepts of space, time, number, etc. would present difficulties of a quite different order if contemporary man were obliged to be the first to acquire them. For the faculty which this contemporary man is to acquire is the combinatory understanding. Logic did not exist among the Atlanteans. But each previously acquired power of the soul must withdraw, in its own form, become submerged beneath the threshold of consciousness if a new one is to be acquired. For example, if the beaver were suddenly to become a thinking being, it would have to change its capacity for intuitively erecting its artful constructions into something else.

The Atlanteans also had for example, the capacity to control the life force in a certain way. They constructed their wonderful machines through this force. But on the other hand, they had nothing of the gift for story-telling which the peoples of the fifth root race possess. There were as yet no myths and fairy tales among them. The life-mastering power of the Atlanteans first appeared among the members of our race under the mask of mythology. In this form it could become the basis for the intellectual activity of our race. The great inventors among us are incarnations of "seers" of the Atlanteans. In their inspirations of genius is manifested what has as its basis something else, something that was like life-producing power in them during their Atlantean incarnation. Our logic, knowledge of nature, technology and so forth, grow from a foundation which was laid in Atlantis. If, for instance, an engineer could transform his combining faculty backward, something would result which was in the power of the Atlantean. All of Roman jurisprudence was the transformed will power of a former time. In this the will as such remained in the background, and instead of itself assuming *forms,* it transformed itself into the forms of thought which are manifested in legal concepts. The esthetic sense of the Greeks is built up on the basis of directly acting forces which among the Atlanteans were manifested in a magnificent breeding of plant and animal forms. In the imagination of Phidias lived something which the Atlantean used directly for the transformation of actual living beings.

What is the relation of the science of the spirit, of theosophy, to the so-called secret sciences?

Secret sciences have always existed. They were cultivated in the so-called mystery schools. Only the one who underwent certain tests could learn something of them. He was always told only as much as was appropriate for his intellectual, spiritual, and moral faculties. This had to be so, for when properly used, the higher insights are the key to a power which must lead to misuse in the hands of the unprepared. Some of the elementary teachings of mystery science have been popularized by the science of the spirit. The reason for this lies in the conditions prevailing in our time. With respect to the development of the understanding, mankind today, in its more advanced members, has progressed to the point where sooner or later it would of itself attain certain conceptions which were previously a part of secret knowledge.

But it would acquire these conceptions in an atrophied, caricatured, and harmful form. Therefore some of those who are initiated into secret knowledge have decided to communicate a part of it to the public. It will thus become possible to measure human advances which take place in the course of cultural development, with the measuring rod of true wisdom. Our knowledge of nature, for example, does lead to ideas about the causes of things. But without a deepening through mystery science these ideas can only be distortions. Our technology is approaching stages of development which can only redound to the good of mankind if the souls of men have been deepened in the sense of the spiritual scientific conception of life. As long as the peoples had nothing of modern knowledge of nature and modern technology, the form was salutary in which the highest teachings were communicated in religious images, in a manner which appealed merely to the emotional. Today mankind needs the same truths in a rational form. The world outlook of the science of the spirit is not an arbitrary development, but results from an insight into the historical fact just mentioned.

However, even today certain parts of secret knowledge can only be communicated to those who undergo the tests of initiation. Only those will be able to make use of the published part who do not limit themselves to an external noting of it, but who really assimilate these things internally and make them the content and the guiding principle of their lives. It is not a matter of mastering the teachings of the science of the spirit with the understanding, but of permeating feelings, emotions, the whole *life* with them. Only through such a permeating does one learn something of their truth. Otherwise they remain something which "one can believe or not believe." When correctly understood, the truths of the science of the spirit will give man a true foundation for his life, will let him recognize his value, his dignity, and his essence, and will give him the highest zest for living. For these truths enlighten him about his connection with the world around him; they show him his highest goals, his true destiny. They do this in a way which corresponds to the demands of the present, so that he need not remain caught in the contradiction between belief and knowledge. One can be a modern scientist and a scientist of the spirit at the same time. But, however, one must be one and the other in a true sense.

xx PREJUDICES ARISING FROM
ALLEGED SCIENCE (1904)

IT IS CERTAINLY TRUE that much in the intellectual life
of the present makes it difficult for one who is seeking
the truth to accept spiritual scientific (theosophical) insights. And what has been said in the essays on the
Lebensfragen der theosophischen Bewegung (Vital Questions of the Theosophical Movement) can be taken as
an indication of the reasons which exist especially for
the conscientious seeker of truth in this respect. Many
statements of the scientist of the spirit must appear entirely fantastic to him who tests them against the certain
conclusions which he feels obliged to draw from what he
has encountered as the facts of the research of natural
science. To this is added the fact that this research can
point to the enormous blessings it has bestowed and
continues to bestow on human progress. What an over-
whelming effect is produced when a personality who
wants to see a view of the world built exclusively on the
results of this research, can utter the proud words: "For
there lies an abyss between these two extreme concep-

tions of life: one for this world alone, the other for
heaven. But up to the present day, traces of a paradise,
of a life of the deceased, of a personal God, have nowhere been found by human science, by that inexorable
science which probes into and dissects everything, which
does not shrink back before any mystery, which explores
heaven beyond the stars of the nebulas, analyzes the in-
finitely small atoms of living cells as well as of chemical
bodies, decomposes the substance of the sun, liquefies
the air, which will soon telegraph by wireless transmission
from one end of the earth to the other, and already today sees through opaque bodies, which introduces navigation under the water and in the air, and opens new
horizons to us through radium and other discoveries;
this science which, after having shown the true relationship of all living beings among themselves and their gradual changes in form, today draws the organ of the human
soul, the brain, into the sphere of its penetrating research." (Prof. August Forel, *Leben und Tod* (Life and
Death) Munich, 1908, page 3) . The certainty with which
one thinks it possible to build on such a basis betrays
itself in the words which Forel joins to the remarks
quoted above: "In proceeding from a monistic conception of life, *which alone takes all scientific facts into account,* we leave the supernatural aside and turn to the
book of nature." Thus, the serious seeker after truth is
confronted by two things which put considerable obstacles in the way of any inkling he may have of
the truth of the communications of the science of the
spirit. If a feeling for such communications lives in him,

even if he also senses their inner well-foundedness by means of a more delicate logic, he can be driven toward the suppression of such impulses when he has to tell himself two things. First of all, the authorities who know the cogency of positive facts consider that everything "supersensible" springs only from day-dreams and unscientific superstition. In the second place, by devoting myself to these transcendental matters, I run the risk of becoming an impractical person of no use in life. For everything which is accomplished in practical life must be firmly rooted in the "ground of reality."

Not all of those who find themselves in such a dilemma will find it easy to work their way through to a realization of how matters really stand with respect to the two points we have cited. If they could do it, with respect to the first point they would, for instance, see the following: The results of the science of the spirit are nowhere in conflict with the factual research of natural science. Everywhere that one looks at the relation of the two in an *unprejudiced* manner, there something quite different becomes apparent for our time. It turns out that this factual research is steering toward a goal which in a by no means very distant future, will bring it into full harmony with what spiritual research ascertains in certain areas from its supersensible sources. From hundreds of cases which could be adduced as proof for this assertion, we shall cite a characteristic one here.

In my lectures on the development of the earth and of mankind, it has been pointed out that the ancestors of the present-day civilized peoples lived in a land-area which at one time was situated in that part of the surface of the earth which today is occupied by a large portion of the Atlantic Ocean. In the essays, *From the Akasha Chronicle,* it is rather the soul-spiritual qualities of these Atlantean ancestors which have been indicated. In oral presentations also has often been described how the earth surface looked in the old Atlantean land. It was said that at that time the air was saturated with water mist vapors. Man lived in the water mist, which in certain regions never lifted to the point where the air was completely clear. Sun and moon could not be seen as they are today, but were surrounded by colored coronas. A distribution of rain and sunshine, such as occurs at present, did not exist at that time. One can clairvoyantly explore this old land; the phenomenon of the rainbow did not exist at that time. It only appeared in the post-Atlantean period. Our ancestors lived in a country of mist. These facts have been ascertained by purely supersensible observation, and it must even be said that the spiritual researcher does best to renounce all deductions based on his knowledge of natural science, for through such deductions his unprejudiced inner sense of spiritual research is easily misled. With such observations one should now compare certain ideas toward which some natural scientists feel themselves impelled at present. Today there are scientists who find themselves forced by facts to assume that at a certain period of its development the earth was enveloped in a cloud mass. They point out that at present also, clouded skies exceed the unclouded, so that life is still to a large extent under the influence of sunlight which is weakened by the formation of clouds, hence one cannot say that life could not have developed under the cloud cover of that Atlantean time. They further point out that those organisms which can be considered among the oldest of the plant world are of a kind which also develop without direct sunlight. Thus, among the forms of this older plant world those desert-type plants which need direct sunlight and dry air, are not present. And also with respect to the animal world, a scientist, Hilgard, has pointed out that the giant eyes of extinct animals, for instance, of the ichtyosaurus, indicate that a dim illumination must have prevailed on the earth in their time. I do not mean to regard such views as not needing correction. They interest the spiritual researcher less through what they state than through the *direction* into which factual research finds itself forced. Even the periodical *Kosmos*, which has a more or less Haeckelian point of view, some time ago published an essay worthy of consideration which, because of certain facts of the plant and animal world, indicated the possibility of a former Atlantean Continent.

If one brought together a greater number of such matters one could easily show how true natural science is moving in a direction which in the future will cause it to join the stream which at present already carries the waters of the springs of spiritual research. It cannot be emphasized too strongly that spiritual research is nowhere in contradiction with the *facts* of natural science. Where its adversaries see such a contradiction, this does not relate to facts, but to the *opinions* which these adversaries have formed, and which they *believe* necessarily result from the facts. But in truth there is not the slightest connection between the opinion of Forel quoted above, for instance, and the facts of the stars of the nebulas, the nature of the cells, the liquefaction of the air, and so forth. This *opinion* represents nothing but a *belief* which many have formed out of a need for believing, which clings to the sensory-real, and which they place *beside* the facts. This belief is very dazzling for present-day man. It entices him to an inner intolerance of a quite special kind. Its adherents are blinded to the point where they consider their own opinion to be the only "scientific" one, and ascribe the views of others merely to prejudice and superstition. Thus it is really strange when one can read the following sentences in a newly-published book on the phenomena of the soul life [Hermann Ebbinghaus, *Abriss der Psychologie* (Outline of Psychology)]: "As a help against the impenetrable darkness of the future and the insuperable might of inimical powers, the soul creates *religion* for itself. As in other experiences involving ignorance or incapacity, under the pressure of uncertainty and the terror of great dangers, ideas as to how help can be found here, are quite naturally forced upon man in the same way in which one thinks of water when in danger from fire, of the helpful comrade in the peril of combat." "In the lower stages of civilization, where man still feels himself to be quite impotent and to be surrounded by sinister dangers at every step, the feeling of fear, and correspondingly, the belief in evil spirits and demons naturally entirely prevail. In higher

stages on the other hand, where a more mature insight
into the interconnection of things and a greater power
over them produce a certain self-confidence and stronger
hopes, a feeling of confidence in invisible powers comes
to the fore and with it the belief in good and benevolent
spirits. But on the whole, both fear and love, side by
side, remain permanently characteristic of the feeling of
man toward his gods, except that their relation to one
another changes according to the circumstances."—"These
are the roots of religion . . . fear and need are its mothers,
and although it is principally perpetuated by authority
once it has come into existence, still it would have died
out long since if it were not constantly being reborn out
of these two."

Everything in these assertions has been shifted and
thrown into disorder, and this disorder is illuminated
from the wrong points of view. Furthermore, he who
maintains this opinion is firm in his conviction that his
opinion must be a generally binding truth. First of all,
the content of religious conceptions is confused with the
nature of religious feelings. The *content* of religious
conceptions is taken from the region of the supersensible
worlds. The religious feeling, for example, fear and love
of the supersensible entities, is made the creator of this
content without further ado, and it is assumed without
hesitation that nothing real corresponds to the religious
conceptions. It is not even considered remotely possible
that there could be a true *experience* of supersensible
worlds, and that the feelings of fear and love then cling
to the reality which is given by this experience, just as

no one thinks of water when in danger from fire, of the helpful comrade in the peril of combat, if he has not known water and comrade previously. In this view, the science of the spirit is declared to be day-dreaming because one makes religious feeling the creator of entities which one simply regards as non-existent. This way of thinking totally lacks the consciousness that it is possible to experience the content of the supersensible world, just as it is possible for the external senses to experience the ordinary world of the senses.

The odd thing that often happens with such views is that they resort to the kind of deduction to support *their* belief which they represent as improper in their adversaries. For example, in the above-mentioned work of Forel the sentence appears, "Do we not live in a way a hundred times truer, warmer, and more interestingly when we base ourselves on the ego, and find ourselves again in the souls of our descendants, rather than in the cold and nebulous fata morgana of a hypothetical heaven among the equally hypothetical songs and trumpet soundings of supposed angels and archangels, which we cannot imagine, and which therefore mean nothing to us." But what has that which "one" finds "warmer," "more interesting," to do with the truth? If it is true that one should not deduce a spiritual life from fear and hope, is it then right to deny this spiritual life because one finds it to be "cold" and "uninteresting"? With respect to those personalities who claim to stand on the "firm ground of scientific facts," the spiritual researcher is in the following position. He says to them, Nothing of what you

produce in the way of such *facts* from geology, paleontology, biology, physiology, and so forth is denied by me.
It is true that many of your assertions are in need of
correction through other facts. But such a correction will
be brought about by natural science itself. Apart from
that, I say "yes" to what you advance. It does not enter
my mind to fight you when you advance facts. But your
facts are only a part of reality. The other part are the
spiritual facts, through which the occurrence of the sensory ones first becomes understandable. These facts are
not hypotheses, not something which "one" cannot imagine, but something *lived* and *experienced* by spiritual
research. What you advance beyond the facts you have
observed is, without your realizing it, nothing other than
the opinion that those spiritual facts cannot exist. As a
matter of fact, you advance nothing as the proof of your
assertion except that such spiritual facts are unknown to
you. From this you deduce that they do not exist and
that those who claim to know something of them are
dreamers and visionaries. The spiritual researcher does
not take even the smallest part of your world from you;
he only adds his own to it. But you are not satisfied
that he should act in this way; you say—although not always clearly—" 'One' must not speak of anything except
of that of which we speak; we demand not only that that
be granted to us of which we know something, but we
require that all that of which we know nothing be declared idle phantasms." The person who wants to have
anything to do with such "logic" cannot be helped for
the time being. With *this* logic he may understand the

sentence: "Our I has formerly lived directly in our human ancestors, and it will continue to live in our direct or indirect descendants." (Forel, *Leben und Tod* (Life and Death), page 21.) Only he should not add, "Science *proves* it," as is done in this work. For in this case science "proves" nothing, but a belief which is chained to the world of the senses sets up the dogma: That of which I can imagine nothing must be considered as delusion; and he who sins against my assertion offends against true science.

The one who knows the development of the human soul finds it quite understandable that men's minds are dazzled for the moment by the enormous progress of natural science and that today they cannot find their way among the forms in which great truths are traditionally transmitted. The science of the spirit gives such forms back to mankind. It shows for example how the Days of Creation of the Bible represent things which are unveiled to the clairvoyant eye.* A mind chained to the world of the senses finds only that the Days of Creation contradict the results of geology and so forth. In understanding the deep truths of these Days of Creation, the science of the spirit is equally far removed from making them evaporate as a mere "poetry of myths," and from employing any kind of allegorical or symbolical methods of explanation. *How* it proceeds is indeed quite unknown to those who still ramble on about the contradiction be-

---

* Compare: Rudolf Steiner, *Die Geheimnisse der biblischen Schopfungsgeschichte* (The Secrets of the Biblical History of Creation), Freiburg i. Br., 1954.

tween these Days of Creation and science. Further, it
must not be thought that spiritual research finds its
knowledge in the Bible. It has its own methods, finds
truths independently of all documents and then recognizes them in the latter. This way is necessary for many
present-day seekers after truth. For they demand a spiritual research which bears within itself the same character as natural science. And only where the nature of
this science of the spirit is not recognized does one become perplexed when it is a matter of protecting the
facts of the supersensible world from opinions which appear to be founded on natural science. Such a state of
mind was even anticipated by a man of warm soul, who
however could not find the supersensible content of the
science of the spirit. Almost eighty years ago this personality, Schleiermacher, wrote to the much younger
Lücke: "When you consider the present state of natural
science, how more and more it assumes the form of an
encompassing account of the universe, what do you then
feel the future will bring, I shall not even say for our
theology, but for our evangelical Christianity? . . . I feel
that we shall have to learn to do without much of what
many are still accustomed to consider as being inseparably
connected with the nature of Christianity. I shall not
even speak of the Six Days' Work, but the *concept of
creation,* as it is usually interpreted . . . How long will
it be able to stand against the power of a world-outlook
formed on the basis of scientific reasonings which nobody can ignore? . . . What is to happen, my dear friend?

I shall not see this time, and can quietly lie down to sleep; but you, my friend, and your contemporaries, what do you intend to do?" (*Theologische Studien und Kritiken von Ullmann und Umbreit* (Theological Studies and Criticism by Ullmann and Umbreit), 1829, page 489). At the basis of this statement lies the opinion that the "scientific *reasonings*" are a necessary result of the facts. If this were so, then "nobody" could ignore them, and he whose feeling draws near the supersensible world can wish that he may be allowed "quietly to lie down to sleep" in the face of the assault of science against the supersensible world. The prediction of Schleiermacher has been realized, insofar as the "scientific reasonings" have established themselves in wide circles. But at the same time, today there exists a possibility of coming to know the supersensible world in just as "scientific" a manner as the interrelationships of sensory facts. The one who familiarizes himself with the science of the spirit in the way this is possible at present, will be preserved from many superstitions by it, and will become able to take the supersensible facts into his conceptual store, thereby divesting himself of the superstition that fear and need have created this supersensible world.

The one who is able to struggle through to this view will no longer be held back by the idea that he might be estranged from reality and practical life by occupying himself with the science of the spirit. He will then realize how the true science of the spirit does not make life poorer, but richer. It will certainly not mislead him into underestimating telephones, railroad technology, and aerial navigation; but in addition he will see many other practical things which remain neglected today, when one believes only in the world of the senses and therefore recognizes only a part of the truth rather than all of it.

INDEX